The Elegant Universe of Albert Einstein

The Elegant Universe of Albert Einstein

THE COLLECTED LECTURES OF THE
ROYAL SOCIETY OF NEW ZEALAND
E=MC² SERIES, BROADCAST ON
NATIONAL RADIO

AWA SCIENCE

First edition published in 2006 by Awa Press, 16 Walter Street,
Wellington, New Zealand

Introduction © Rebecca Priestley 2006
Essays © individual authors as credited
Awa Press acknowledges the generous assistance of the Science Photo Library
in providing many of the illustrations in this book.

The right of the authors to be identified as the author of this work in terms of
Section 96 of the Copyright Act 1994 is hereby asserted.

This book is sold subject to the condition that it shall not, by way of trade
or otherwise, be lent, resold, hired out or otherwise circulated without the
publisher's prior consent in any form of binding or cover other than that in
which it is published and without a similar condition including this condition
being imposed on the subsequent purchaser.

National Library of New Zealand Cataloguing-in-Publication Data
The elegant universe of Albert Einstein : leading New Zealand scientists and
historians on the greatest revolution in modern science / Tom Barnes ... [et al.] ;
introduction by Rebecca Priestley.
(Awa science)
Includes index.
ISBN: 0-9582629-2-6
1. Physics. I. Barnes, T. H. (Thomas Heinrich), 1953-
II. Series.
530—dc 22

Published with the generous support of the Charles Fleming Fund; the Division
of Humanities, University of Otago; and the Royal Society of New Zealand.

Designed and typeset by The Letterheads.
This book is typeset in Mercury Text, Grade Two & Maple.
Printed by Printlink, Wellington

www.awapress.com

Contents

Introduction · *Rebecca Priestley* 1
A Short History of the Universe · *Matt Visser* 13
Discovering the Age of the Earth · *Hamish Campbell* . . . 29
Einstein and the Eternal
   Railway Carriage · *Lesley Hall & Richard Hall* 49
Schrödinger's Cat · *Tom Barnes* 73
Journey to the Heart of Matter · *Paul Callaghan* 99
The Unconquered Sun · *Robert Hannah* 119
Galileo's Dilemma · *John Stenhouse* 145
Further Reading 164
Index 167

Introduction
—*Rebecca Priestley*

MENTION ALBERT EINSTEIN to a New Zealander and chances are they'll think of E=mc² and picture a crazy-haired man with a bushy moustache and one of the most recognised faces in the world. The truth is that E=mc², a monumental implication of Einstein's special theory of relativity, was just one of Einstein's many achievements. The man whom *Time* magazine named the most important person of the twentieth century, and whose brain alone has been the subject of several books, in 1905 laid the foundations for two remarkable new concepts in physics, relativity and quantum theory. While he continued working and contributing to physics until his death in the United States in 1955, it is his work in 1905 that stands out, and the year has come to be known as Einstein's *annus mirabilis*, his miraculous year.

In this year of extraordinary scientific creativity, Albert Einstein was 26. He had been born in Germany in 1879, to a family of secular Jews. Little Albert, now widely considered one of the greatest geniuses of the twentieth century, showed no immediate signs of cleverness. In fact he was so slow to learn to speak — he didn't begin to talk until after his second birthday — that his parents consulted a physician, fearing that he was mentally deficient. Once he did start to speak, however, he proved them wrong; he spoke in carefully considered sentences, which he would often practise in a whisper before speaking out loud.

Einstein began his education at the local Catholic school in Munich, and continued at the Luitpold Gymnasium. At school he was an independent and sometimes disruptive student, choosing to concentrate his efforts on his favoured subjects of physics, mathematics and philosophy and largely ignoring subjects – such as foreign languages – in which he had little interest or aptitude. While his schooling was important in introducing him to subjects like physics and mathematics, he found more inspiration outside the classroom than in it. In his *Autobiographical Notes*, Einstein described experiencing 'a miracle' when, as a sick child of five, he was given a compass to play with by his father. Einstein was transfixed by the new toy, twisting and turning it, trying to fathom the invisible force that kept the magnetic needle pointing north.

Einstein's father and uncle owned and ran a moderate-sized factory, producing electrical equipment for municipal power stations and lighting systems. In 1894, in search of better business opportunities, the family moved from Munich to Milan in northern Italy, leaving 16-year-old Einstein behind to finish his education. He was unhappy with this arrangement, and six months later surprised his parents by appearing at their new home in Milan, waving a doctor's certificate attesting to a nervous disorder and announcing that he had left school. Einstein had disliked the authoritarianism of his German school, and anticipated with horror the compulsory military training that would soon be his fate. After leaving Germany, he renounced his German citizenship and swore never to return. (He would later break this vow, regaining German citizenship to live and work in Berlin as Professor of Physics at Berlin University from 1914 until Adolf Hitler's fascist policies and Germany's growing

Introduction

anti-Semitism drove him to the United States in 1933.)

Einstein completed his schooling in Switzerland, gaining entrance to the Swiss Federal Polytechnic in Zürich and registering for a course that would qualify him as a specialised teacher of physics and mathematics. He did well in his classes, while also enjoying sailing on the lake, playing and listening to music, and sitting in Swiss *Kaffeehauser* having passionate discussions about physics and the problems of the world with his fellow students. With his dark good looks (he didn't always have that unkempt white mop) and talent for the violin, he was popular with his female acquaintances – including his one female classmate Mileva Marić, a Serbian, with whom he developed a relationship.

Einstein graduated from the Polytechnic in 1900, qualified to teach high-school mathematics and physics, but was disappointed not to gain an assistant's position in the Polytechnic's physics department. He considered his options. The Polytechnic did not offer doctoral degrees, but it had an agreement with the University of Zürich whereby graduates could submit a dissertation to the university for consideration for a PhD. Neither Einstein nor his now fiancée Marić (his parents did not approve of the match and the couple delayed their marriage) had shone in their final exams – he had passed, she had not – and as no teaching positions were on offer Einstein paid to use the Polytechnic's laboratory for research into heat and electricity, with a view to submitting work towards a PhD. Under financial and parental pressure, Einstein soon abandoned this plan and in 1901 he returned to Milan. He eventually found a temporary job teaching in a technical school and moved to Winterthur in Switzerland.

In 1902 Einstein moved again, this time to Bern, where he had applied for a job with the Swiss Patent Office. While waiting for his application to be processed, he worked as a tutor and engaged in lively discourse with his students. Together they read and discussed a wide range of works by scientists and philosophers, becoming fast friends in the process. The same year, Marić, now living with her parents in what was then Hungary, gave birth to their child, a girl they called Lieserl. Einstein's tutoring was more a hobby than a job – it paid only enough to keep him in tobacco – so he was relieved to be granted the patent office job in 1902. The next year, after his father gave permission on his deathbed, Einstein and Marić married in Bern. Little Lieserl's fate is unknown – it was unacceptable in turn-of-the-century Europe for a woman to bring up an illegitimate child, or for an unmarried couple to live together, and it is likely she was adopted or died in infancy – but in 1904 the first of the couple's two sons, Hans Albert, was born.

By 1905, then, Albert Einstein was a university graduate, a husband and a father, working as a technical assistant in the Swiss Patent Office in Bern. Since graduating he had continued his own research and between 1900 and 1904 had had several scientific papers published. Marić, who had failed her polytechnic examinations a second time, had given up her hopes of teaching to look after her husband and new baby. Einstein, nonetheless, was charged with tending to the child often enough to master the art of rocking the cradle with his foot while immersed in a book, and on walks demonstrated his lateral thinking by designating the baby carriage as a mobile desk, stopping occasionally to jot down an inspired thought or scribble out an equation.

Introduction

Albert Einstein. AMERICAN INSTITUTE OF PHYSICS/SPL

By the time Einstein's greatest ideas came to fruition, the scientific world was already reeling from the new physics coming out of Europe. The German physicist Wilhelm Conrad Röntgen's 1896 discovery of X-rays had been closely followed by French physicist Henri Becquerel's discovery of radiation, and Marie and Pierre Curie's discovery of the radioactive elements polonium and radium. These advances, along with the English physicist J. J. Thomson's discovery of the electron, had revealed the existence of a subatomic world. With this radical departure from classical physics – which held that the atom was indivisible – a new age had begun. New Zealander Ernest Rutherford, inspired by the new discoveries, put aside his research into electromagnetism at Cambridge University and began working on X-rays with J. J. Thomson, who was his professor. At the same time, the German physicist Max Planck theorised that light, heat and other forms of radiation were emitted in discrete packets of energy, or 'quanta'. He put this theory forward to allow for ease of a mathematical equation, and did not initially believe that it corresponded with reality.

In 1905, his *annus mirabilis*, as well as holding down a fulltime job Einstein completed his PhD dissertation and published a series of ground-breaking papers in the German physics journal *Annalen der Physik*. His work, along with the revelations in physics that had begun in 1896, blew apart ideas that had held sway since Isaac Newton had initiated the last great revolution in physics in the 1600s.

Einstein's most radical paper, *On a heuristic point of view concerning the production and transformation of light,* challenged the accepted wisdom that light was an electromagnetic wave. Inspired by Max Planck's recent work, Einstein showed that light was indeed made up of discrete packets of energy, or quanta, despite also having wave-like properties. With its concept of wave-particle duality – that light sometimes behaved like a particle and sometimes behaved like a wave – this paper was so revolutionary that it took 16 years for its significance to be fully realised and for Einstein to be awarded the Nobel Prize in physics for it. The paper, along with Planck's earlier work, laid the groundwork for what is now called quantum theory, a cornerstone of modern physics, which states that energy can be absorbed or radiated only in discrete values.

A second paper, *On the movement of small particles suspended in liquids at rest required by the molecular-kinetic theory of heat*, concerned Brownian motion, the irregular and unexplained movement of microscopic particles, such as pollen grains, suspended in water. Einstein proved that the incessant jiggling of

Introduction

the tiny particles was caused by collisions between the suspended particles and the much, much smaller water molecules. This was more radical than it now sounds, for in 1905 many scientists doubted the actual existence of molecules and atoms, using them only as terms to describe scientific concepts. Einstein showed that by measuring the distance the dancing particles travelled over time, it would be possible to calculate the number of molecules in a given volume of liquid or gas, thereby giving an indication of the size of a molecule.

His doctoral dissertation, on a related topic, was inspired by the sight of sugar dissolving in water. It too was completed in 1905, and was published the following year under the title *A new determination of molecular dimensions*. In this paper Einstein calculated the size of a sugar molecule as one-twentieth of a millionth of an inch across. Einstein's 1901 attempt at a doctoral dissertation had not been well-received, and he had withdrawn it (though he was poor at the time and this was perhaps partly for the 230-franc refund). His 1905 dissertation was also initially rejected – criticised for being too short. He added a single sentence, the dissertation was accepted, and Albert Einstein became Dr Einstein.

Einstein's most famous equation, $E=mc^2$, came from one of two 1905 papers that introduced what is now called Einstein's special theory of relativity, in which he showed that the concept of universal or absolute time must be abandoned. The first paper, *On the electrodynamics of moving bodies*, concerned the nature of space and time. Since he was a teenager Einstein had entertained himself by wondering what he would see if he were to travel alongside a beam of light, which by the end of the nineteenth century was known to travel at approximately

186,000 miles per second (now more commonly expressed as 300,000 kilometres per second). In this paper, he abandoned previously held assumptions and determined that if the speed of light is constant, then measurements of space and time must be *relative* – that is, they depend on the location and the movement of the person observing them. Einstein's biographer Denis Brian described the paper as 'strangely free of footnotes or references, as if the inspiration had indeed come, if not from God, from some other-worldly source'. The second short paper on relativity, *Does the inertia of a body depend on its energy content?*, yielded the equation $E=mc^2$ and the remarkable summation that mass could be turned into energy and energy could be turned into mass.

While Einstein's work was not directly related to that of Ernest Rutherford, the two men were each at the forefront of theoretical physics. In a 1903 paper called *Radioactive change*, Rutherford and his colleague Frederick Soddy had concluded that 'the energy latent in the atom must be enormous'. Rutherford and Soddy speculated about the release of energy from the atom and the possibility of atomic weapons, with Rutherford jokingly suggesting that 'some fool in a laboratory might blow up the universe unawares'. Einstein's paper showed that when an atom emitted energy in the form of radiation, its mass was reduced by a proportional amount, with the equation $E=mc^2$ putting a figure to the enormous amount of energy inherent even in one tiny atom.

In 1939 this equation was used by the Austrian physicist Lise Meitner to calculate the potential release of energy from the splitting, or fission, of the uranium atom, which she and her colleague Otto Hahn had recently achieved. Her conclu-

Introduction

sion was published in *Nature*, and a global search began for a way to harness this energy. With war imminent, the focus was on using the energy for a bomb. Einstein's other link with the atomic bomb soon followed. By 1939 Einstein was world famous, a Nobel Prize-winning scientist (long divorced from Marić and now a widower after the death of his second wife), living and working in the United States. As a representative of a group of German scientists who had escaped to the United States, he wrote to President Franklin Roosevelt, warning of the progress in uranium research in Germany and urging the president to consider having the US undertake the same, suggesting that the fission of uranium could be used to create an atomic explosion.

But in 1905 neither Einstein nor anyone else had any idea of the enormous implications of his work in physics. And in New Zealand, far from the excitement in Europe, gentlemen scientists focusing on natural history and geology dominated science. In the *Transactions and Proceedings of the New Zealand Institute,* the journal of the state-coordinated network of scientific societies, the occasional articles on physics were usually relegated to the 'miscellaneous' category, where they sat alongside articles on archaeology, ethnology and literature. In the New Zealand universities, physics was taught by either junior staff or by chemistry professors, with the country's first physics professor not appointed until 1909, at Victoria University College. Young physics graduates such as Ernest Rutherford, who had himself left the country ten years earlier, were forced to move to Europe to continue their studies.

The evolving revolution in physics, however, did begin to attract the attention of New Zealand scientists and was much discussed at regional meetings of the philosophical societies that collectively made up the New Zealand Institute (now the Royal Society of New Zealand). It took time for Albert Einstein's 1905 achievements to be fully appreciated by the scientific world, let alone the general public, but the public was hungry for science, especially when it could be demonstrated as excitingly as the new X-rays and radiation technologies. Demonstrations of 'skiagraphs', or X-ray photographs, were enthusiastically received, with no shortage of volunteers willing to have the newly discovered X-radiation aimed at their hands, heads or limbs in order to gain a picture of their previously hidden skeletons. Experiments involving the expensive and hard-to-obtain radium were also popular. In 1905, the year Einstein published his remarkable papers, Ernest Rutherford, by then Professor of Physics at McGill University in Canada, visited New Zealand and lectured to crowded halls on 'The wonders of radium', entertaining the crowds with experiments using a sample of radium he had obtained in Europe.

One hundred years later, New Zealanders would again flock to hear scientists lecture on the wonders of physics. To celebrate the centenary of Einstein's *annus mirabilis*, UNESCO had declared 2005 the International Year of Physics. As part of a series of events around the country, the Royal Society of New Zealand organised a lecture series in which invited scientists spoke about some of the big ideas in the physical sciences, with their talks broadcast by Radio New Zealand. While some of the $E=mc^2$ lectures collected in this book focus on Einstein's work and its repercussions for twentieth century physics, others reach back into the past to present some of the equally thrilling discoveries that preceded it, and showed the way. Perhaps most excitingly, the writers also look to the future – to the birth of new stars,

Introduction

the explosion of new ideas – but also to an uncertain fate for humanity. The scientific discoveries recounted here are enlightening and stimulating, but many scientists believe we are still in the early stages of learning about the nature and potential of our universe. The most exciting discoveries in science may yet await us.

A Short History of the Universe
—*Matt Visser*

TWO THOUSAND AND FIVE, the International Year of Physics, marked a century since Albert Einstein published his first papers on special relativity, the photoelectric effect (one of the foundations of quantum physics) and Brownian motion (which led to precisely measuring the size of atoms). Looking back, it is difficult now to realise just how stunning these articles were at the time. They challenged mankind's perception of reality, and so laid the foundation for most twentieth century physics and technology.

The year 2005 also marked 90 years since the publication of Einstein's theory of general relativity. The general relativity of 1915, as opposed to the special relativity of 1905, is Einstein's theory of gravity – the theory that leads physicists, astronomers and mathematicians to talk about such notions as black holes and the 'big bang'. It is the foundation stone of modern cosmology. And if you are willing to indulge in a little speculation, it is Einstein's general relativity that leads to the ideas of wormholes in space-time, and even time travel. But let us leave the speculation to the science fiction community and Hollyweird, and talk about things that are definitely mainstream, that have a tremendous amount of backing in the form of direct empirical data and observation.

By the late 1700s it was already clear to astronomers and physicists that the stars, whatever their source of energy, had limited lifetimes: eventually they would burn out and go dark.

So even then, more than 200 years ago, there was serious discussion of how stars and their accompanying stellar systems – their planets, asteroids and comets – might form, grow, age and die. Much of the detail of that discussion is now irrelevant: back then physicists and astronomers had no concept of atomic or nuclear structure, so their estimates of a star's lifetime are, to modern ears, ludicrously off-base. It was only with the revolution in physics in the twentieth century – a revolution largely, but not entirely, attributable to Albert Einstein – that scientists became able to produce accurate models of how the stars are put together, how they shine, and how long they will last. These days, when your child wanders up to you and says, 'Dear parental unit, why does the sun shine?' you know the answer: the sun shines thanks to the thermonuclear burning of hydrogen to helium deep in its core. This thermonuclear burning (for which no oxygen is required, because it is not a chemical process) heats the core to several millions of degrees Celsius.

Now, the general framework in which present-day physicists and astronomers work when dealing with questions of the origins and ultimate fate of the universe is that of 'big bang' cosmology. Cosmology is the study of the universe as a whole, and before Einstein developed his theory of general relativity cosmology was more a part of philosophy and religion than of science. Einstein's theory changed all that: we could begin to formulate cosmological questions in a precise mathematical framework that was both logically consistent and subject to empirical testing.

General relativity, stripped to its essentials, is the dynamics of geometry, sometimes called geometrodynamics. Slightly modifying the words of the physicist responsible for coining the phrase 'black hole', American John Wheeler, 'the geometry of space and time tells matter how to move, and the presence of matter tells space and time how to warp and twist.' This arena of space and time, in terms of which general relativity is formulated, is lumped together by mathematicians and physicists into a unified whole: space-time.

The key mystery of Einstein's general relativity is this: the geometry of the universe is not that of Euclid. Indeed, the classical Euclidean geometry developed by the ancient Greeks – that of triangles, spheres, cones, the Pythagoras theorem, and in particular Euclid's famous five axioms – is only an approximation of the way the universe really functions. Euclid's geometry works perfectly well in flat space, but the presence of matter warps both space and time so they are no longer flat.

In reasonably small chunks of space-time, Euclidean geometry is a truly excellent approximation. If we try to look very closely at an extremely small patch of space-time, we expect that eventually quantum weirdness may come into play, and that quantum physics may modify Euclidean geometry. But if such quantum-induced changes in the geometry of our universe do occur, we expect that to happen only at extremely small distances, about a million million million times smaller than the nucleus of an atom, and well beyond our present-day experimental and observational probes.

On the other hand, at large distances we definitely begin to see deviations from Euclid's idealised geometry, and these deviations are well within our ability to test experimentally and observationally. They already show up in the Global Positioning System, whose designers, the United States Air Force, had to take Einstein's general relativity into account when setting it up.

Further out in our solar system, effects such as the bending of starlight by the sun, the slowing of clocks deep in a gravitational field, and the time delay that radio signals suffer when crossing a gravitational field, are all aspects of subtle non-Euclidean geometry in our planetary neighbourhood – and all features of general relativity that we can, and do, regularly test with astronomical observations.

It is once we move beyond our own solar system, and even beyond our own galaxy, the Milky Way, that things get really interesting. As we look further out into the cosmos, we see something that at first stunned the astronomers of the 1930s: other galaxies are moving away from us. And the further we look, the faster they are moving. This is the Hubble flow, the recession of the galaxies resulting from the expansion of the universe. If we naïvely project this Hubble flow backwards in time, we find that some ten thousand million years ago the galaxies must have been sitting on top of each other. This is the observational foundation stone underlying 'big bang' cosmology. And when you get a little bit fancier and actually solve the Einstein equations, instead of making an eyeball estimate based on the present-day Hubble flow, the age of the universe comes out as approximately 14 thousand million years.

As best we can tell, the history of our universe goes like this. Some 14 thousand million years ago, give or take the odd thousand million years, the universe is in an extremely hot, extremely dense phase of matter and energy. This extremely hot, extremely dense fireball is expanding. It is too hot and dense for objects such as stars and planets to exist, and even too hot and dense for individual atoms, or even nuclei of atoms, to exist. It is a soup of elementary particles called a quark-gluon plasma.

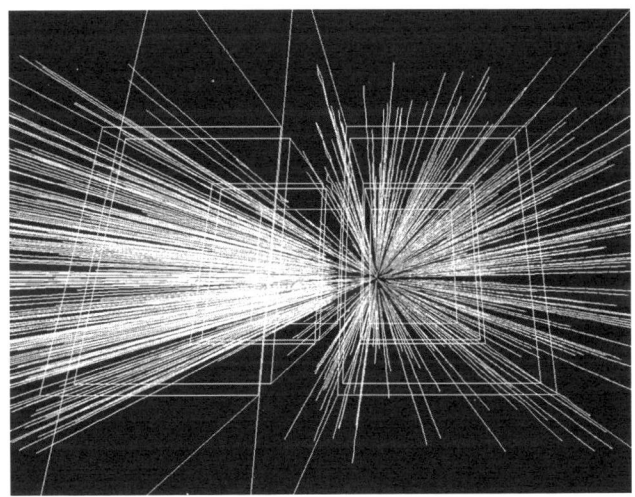

Quark-gluon plasma particle tracks, as constructed by computer. A quark-gluon plasma –a soup of elementary particles – is believed to have emerged from the 'big bang' and contained the fundamental building blocks of matter in the universe. CERN/SPL

As the universe expands and the temperature drops, the quarks first condense to form protons and neutrons. Later, some fraction of these protons and neutrons burn – via thermonuclear fusion – to form helium. This is the epoch of cosmological helium production, which took place long before the stars formed, but still accounts for the majority of the helium in the universe today.

After helium is produced, the fireball consists of a plasma of protons, helium nuclei, electrons and photons (quanta of light), with a few relics such as neutrinos (particles associated with the weak interactions) and gravitons (hypothetical quanta of gravity) thrown in for good measure. Heavier nuclei such as carbon and oxygen have not yet had a chance to form. Because the fireball is still a plasma it contains positive charges, in the form of nuclei, and negative charges, in the form of electrons, that are free to move about – and so it conducts electricity. And because light itself is ultimately electromagnetic radiation and is 'short-circuited' by any conductor, the plasma is opaque. In more picturesque language, the photons – the quanta of light – are being scattered back and forth, all over the place. If you were magically transported back to that time, you would not be able to see through the dense glowing muck. (You would also be quickly crisped to a cinder, but that's another matter.)

As the universe continues to expand the temperature drops, and eventually the electrons start to pair up with the protons and helium nuclei to form hydrogen atoms and helium atoms. This epoch is traditionally referred to as that of 'recombination', although it should really be called 'first combination', or even 'first atom formation'.

Soon the plasma has pretty much disappeared, converted into neutral, non-conducting atoms. And since these atoms are hydrogen and helium gas, they are transparent. The photons are now essentially free to fly across the universe, having largely 'decoupled' from the matter in the cosmos. Today, astronomers and physicists regularly detect photons from this epoch in their experiments. The primordial photons, which have been freely streaming through the universe since the fireball became transparent, are known as the cosmic microwave background or cosmic background radiation, or most often just CMB or CBR.

At decoupling, the cosmic fireball was still very hot – about 3000 degrees Celsius. The photons had a characteristic distribution of energies related to this temperature; in technical

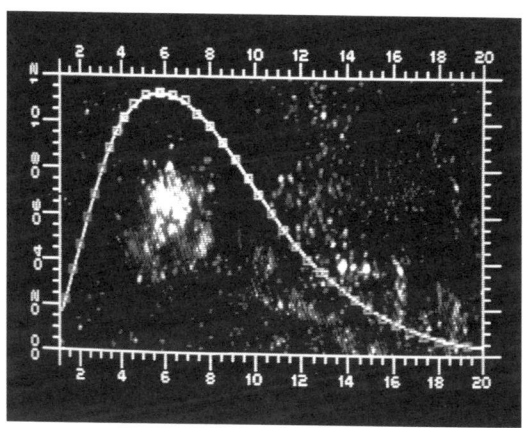

The spectrum of cosmic background radiation obtained by the Cosmic Background Explorer satellite (COBE) in 1990. The radiation is essentially the heat of the 'big bang', diluted with the expansion of the universe to a faint microwave glow. MEHAU KULYK/SPL

parlance, the distribution of energies was a Planck spectrum (often called a black body spectrum), with a characteristic hump at a place corresponding to the 3000 degrees Celsius.

Of course, when we look at the cosmic microwave background today, we do not measure a temperature of 3000 degrees Celsius, but only about 3 degrees above absolute zero. (Make that 2.7 Kelvin to be more precise – about minus 270 degrees Celsius.) The reason is that as the universe expands, so does the wavelength of the light. In fact, the wavelength of the light and the size of the universe grow in lockstep. Since the physical speed of light is constant, the frequency of the light – the number of beats per second in the electromagnetic field – decreases as the universe expands. Light that at decoupling was concentrated in the extreme ultraviolet – and we are talking serious

Arno Penzias and Robert Wilson, the US radio astronomers who unexpectedly discovered CBR in 1963, photographed at Bell Laboratories in 1978, the year they won the Nobel Prize. PHYSICS TODAY COLLECTION/ AMERICAN INSTITUTE OF PHYSICS/SPL

sunburn territory – becomes blue, green, red and infrared, until it is concentrated in the microwave radio band, which is where radio astronomers Arno Penzias and Robert Wilson discovered the cosmic background radiation in 1963: it manifested as an unwanted and unavoidable hiss in their microwave antennae. This discovery, at Bell Laboratories in New Jersey, is often viewed as having clinched the big bang theory.

Between decoupling, when the universe became transparent, and today, the wavelength of these primordial photons has been stretched by a factor of about 1000, the frequency reduced by a factor of about 1000, and the temperature of the light from the fireball reduced by a factor of about 1000. This means that the size of the universe has expanded by a factor of about 1000 in all three directions, and the volume by a factor of about one thousand million.

A Short History of the Universe

AFTER DECOUPLING, the temperature of the fireball continues to drop. It's still a relatively long time until the first stars begin to form. The temperature has to drop to about 30 degrees Celsius, remarkably close to room temperature, before baby stars have even a chance of getting past the first hurdle. But what is a 'baby star'? Stripped to its essentials, the fireball contains small fluctuations in density – places where the fireball is a little more concentrated than average, and places where it is a little less concentrated. The regions that are a little more concentrated attract yet more matter to themselves via gravity, while the regions with less matter than average have the remaining matter sucked out of them by the gravity of the over-endowed regions. The result is positively biblical:

To him that hath, more shall be given; and he that hath not, from him shall be taken even what little he hath.

This process, continuing over aeons, leads to the formation of stars, planets and even the galaxies themselves.

The very early stages, the irregularities that exist at the time of decoupling, leave a measurable imprint on the cosmic background radiation. The temperature of this radiation is not exactly 2.7 Kelvin in all directions: there are small fluctuations, about one part in a million, scattered over the sky. The detectors of today's physicists and astronomers are good enough to measure these tiny temperature differences, and to provide a sky map of temperature fluctuations measured now. This corresponds to a sky map of temperature fluctuations at decoupling, which translates to a sky map of *density* fluctuations at decoupling and so, finally, to a map that tells us which parts of the sky had a little more matter than average and which had a little less at the epoch of decoupling, about 13 thousand million years ago.

These tiny variations, which we can observe directly in the cosmic background radiation, are the seeds that will grow up to be the stars, galaxies and, after a bit of recycling, the planets – including our own. But in our discussion we are nowhere near planets forming just yet; we have just got to the first stars forming when the temperature of the fireball drops to about 300 Kelvin, which was about 12 thousand million years ago. These very first stars consist of hydrogen and helium, and nothing else. The heavier elements have not yet had a chance to form.

It is only as the first generation of stars ages and begins to die that the first heavy elements are formed by thermonuclear fusion – and I emphasise that, to an astronomer, anything heavier than the very lightest elements, hydrogen and helium, is a heavy element. As the first generation of stars age, their cores first burn hydrogen to helium, then helium to carbon, oxygen, and yet heavier nuclei, eventually moving all the way down the curve of binding energy to reach iron 56, a particular isotope of iron that is the most tightly bound of all nuclei, with the least amount of free energy. This is the end-point of the thermonuclear fusion process: there is no more fusion energy available, and iron 56 is completely stable against nuclear decay.

Now, some of these first-generation stars will just burn out and fade away, but others will tap into their gravitational potential energy, letting their central cores shrink while their outer layers expand. Some stars will shed their outer layers in a relatively gentle manner and quietly burn to a cinder but others, perhaps the majority, will undergo violent supernova explosions as their cores radically destabilise. Supernova explosions do two things: they spread the guts of burnt-out stars all over creation; and they convert some gravitational potential energy into heat and light, pushing some of those iron 56 nuclei up the other side of the curve of binding energy in an inverse nuclear fission process that absorbs energy, but (finally) provides the universe with elements all the way up to uranium.

The shock waves from the supernova explosions that spread the heavy elements around also help collapse some of the remaining lumpy dust clouds, leading to the birth of second-generation stars, which are now typically clumped into galaxies. When these stars begin to form, we have a much more interesting mix of elements to work with. Hydrogen and helium still predominate, but there is a small percentage of heavier elements. These modify the thermonuclear dynamics and provide material for the formation of rocky planets and asteroids.

It is now about eleven thousand million years ago. The first generation of stars were typically much larger than most of the stars we now see near us, and they burned very bright and died very quickly, lasting only one thousand million years or so. Second-generation stars are, by and large, more sedate and live longer on average, about six thousand million years, partly because they are typically smaller and partly because the heavy elements produced in the first generation act as a moderator, slowing down the headlong rush to burnout. But when the second generation does burn out, some of these stars also undergo supernova explosions, spreading yet more heavy elements around the universe in preparation for the third generation.

As with human genealogy, this talk of 'generations' is to some extent convention, since stars are certainly being born and dying continuously between the generations. Nevertheless, it is a useful convention, with our own sun normally being ascribed to the third generation. Because we are 'on the spot' we have many more direct means of measuring age, and can pin down the birth of our sun and the condensation of planet Earth out of nebular dust to a mere four and a half thousand million years ago. And we can be pretty confident we have the timing right to within a hundred million years or so, considerably better than the thousand-million-year ambiguities our story has had to deal with so far.

Note that we have had to get up to three generations of star formation to even reach the formation of our own planet, and that all the heavy nuclei we see around us – the carbon, oxygen, nitrogen, silicon, rocks and metals – were born in the cores of earlier generations of stars and then released into the universe via supernova explosions. (The only material on Earth that can be traced back, essentially unaltered, to the quark-gluon phase transition is the hydrogen in water and organic materials. Even the helium, of which there is comparatively little, dates back at best to the cosmological helium-burning episode and much of it is much younger than that.)

Hence, stars are the crucible of life in at least two senses: the star that is our sun provides the heat and light on which our biosphere runs, and previous generations of stars provided the raw material that makes up planet Earth.

Although our sun has a finite lifetime, our best estimates are that it still has about two thousand million years left, so we should not feel any immediate concerns in this regard. There are plenty of other problems that can knock our civilisation for a loop on a much shorter timescale. In the long run, though, it

A Short History of the Universe

will die, in the process destroying our planet. Near the end of its lifetime the sun will swell to a red giant phase, engulfing and melting the Earth, and indeed all the inner planets, before finally recontracting to a cold, burnt-out cinder. (Our sun is too small for it to undergo a supernova explosion, although an ordinary nova is not precluded.)

There will certainly be a fourth generation of stars. Indeed in our telescopes we can see star birth occurring right now. There are dust clouds in the Eagle nebula just starting to shine with inner thermonuclear light as the dust collapses, condenses, and heats up due to compression – just as a bicycle pump heats up after vigorously compressing air. Once the dust gets up to about one million degrees Celsius, thermonuclear reactions start and the star begins to shine with nuclear light. This ignition takes a very short time: it may be as little as 100 years between the final collapse and the emergence of a fully fledged star. The formation of planetary systems seems to take somewhat longer, but still an amazingly short time by geological standards. Evidence from meteorites found in our own solar system indicates that the actual planetary formation phase might take as little as 100,000 years, a blink in the eye compared to the four and a half thousand million years our sun has been shining, or the 14 thousand million years the universe has existed.

Despite such births our solar system will eventually die, as will all stars and their planetary systems. Eventually, as new generations of stars eat up more of the hydrogen, star formation will cease altogether and the universe itself will die, an event poetically referred to as the 'heat death', or more accurately the 'entropy death'.

25

As long as there are temperature differences – or, more technically, entropy gradients – in the universe, then life (although possibly not life as we know it) can survive by feeding off the entropy gradient. In a sense, this is what the entire biosphere of the Earth currently does. It imports lots of high-quality, low-entropy sunlight which arrives via the visible part of the spectrum, and exports lots of low-quality, high-entropy waste heat in the form of infrared radiation. It is ultimately the temperature difference between our sun and the night sky that permits life to exist on Earth. (Our sun has a surface temperature of thousands of degrees Celsius, while the night sky has an overall temperature of 3 Kelvin, or minus 270 degrees Celsius – remember that cosmic background radiation I mentioned before?)

The same basic physics affects each one of us directly and personally. You, as an individual human being, survive by importing high-quality low-entropy material into your body. It's called food. You also export an equal mass of low-quality high-entropy material back into the environment, and I trust I do not have to go into explicit details on that. As long as you export more entropy than you import, your body can use that entropy difference to do interesting things. That's called being alive. In the absence of an entropy difference between imported and exported material, you will die. Apply the same principles to the entire universe, and once the temperature differences in the universe shrink to zero the entire universe will die.

LET US TEMPORARILY AGREE to ignore our petty family squabbles here on Earth, be they political, economic, ethnic or religious. Are there things 'out there' that we, as a species, should be worried about? One obvious candidate is asteroid or cometary impact. We know that large asteroids and comets periodically hit the Earth, and we have good reason to believe that these impacts have had something to do with major extinction events. Certainly, something big hit the Earth 67 million years ago, suspiciously close to when the dinosaurs died out. And it wasn't just the dinosaurs: every land-dwelling animal heavier than about 20 kilograms seems to have snuffed it.

Several other impacts have also occurred suspiciously close to extinction events, and at least one was even more dramatic than the dinosaur die-off. Of course I'm talking millions of years ago, but try this recent incident for size. Just eleven years ago, in July 1994, the comet Shoemaker-Levy collided with the planet Jupiter. We are not talking about a few meteorites, or a bit of comet dust drifting gently down through the Jovian atmosphere. We are talking about enormous, continent-wrecking slabs of methane ice screaming out of the sky at 30 kilometres per second, with impacts releasing energy in the 100-gigatonne range – significantly more than the combined energy release of all the nuclear weapons mankind has ever built – and punching holes the size of Europe in Jupiter's cloud cover.

The fact that in the relatively short time we have been able to make detailed surveys of the rest of our solar system we have already seen one such planet-wrecking impact should make us stop and think a little. If there were any life down in Jupiter's atmosphere in 1994, the arrival of Shoemaker-Levy certainly put paid to it.

So what is to be done? In the short run we need to get our own house in order, both politically and ecologically. But in doing so we should not ignore the universe around us: there is danger and the potential for 'interesting times' out there as well. We should certainly keep an eye on our backyard, the solar system. Remember that we are quite literally living on the 'third rock from the sun': the phrase is not merely the name of a Hollywood sitcom. Make that the third *big* rock; there's quite a bit of rubble left over from the formation of the solar system, which is a good reason for not just focusing on our planetary squabbles but keeping one eye on the sky.

In the longer run, we might want to think about moving house before the sun burns out, but that is some way off. Even further into the future, if the thought of the entropy death concerns you too much you might want to think about moving to another universe. Meanwhile, as Albert Einstein famously said: 'Keep an open mind; but not so open that your brains fall out.'

Matt Visser *was born in Wellington and studied at Victoria University before undertaking his PhD at the University of California at Berkeley. He remained in the United States for 24 years, with post-doctoral stints at the University of Southern California and at Los Alamos National Laboratory before he moved to Washington University in Saint Louis. In 2003 he returned to Victoria University to take up a position in the mathematics department. His research focuses on general relativity, black holes and cosmology.*

Discovering the Age of the Earth
—*Hamish Campbell*

IN MY ROLES AS A GEOLOGIST at Te Papa Tongarewa, the Museum of New Zealand, and a palaeontologist at GNS Science (formerly the Institute of Geological and Nuclear Sciences), the most common questions I am asked are, 'How do you know it is that old?' and 'How do you date things?' To answer these questions, I want to start with the Victorians, for it was they who set the scene: almost all the fundamental developments of modern ideas in science can be traced to scientists who were either born or died during the lifetime of Queen Victoria.

First, some background. In 1644, the vice-chancellor of Cambridge University, a man called John Lightfoot, declared that, from his interpretation of the Bible, the Earth had been created at nightfall on the Sunday preceding the autumn equinox in 3929 BC. In 1650, an Archbishop James Ussher, based on *his* interpretation of the Bible, came up with a different year. The Earth, he said, had actually been formed at nightfall preceding Sunday, 23 October 4004 BC. Ussher had weighty credentials – he was not only head of the Anglican Church of Ireland but also a former Professor of Theology at Trinity College, Dublin – and his date would prove the more popular.

This mantra stood not just for decades but for centuries. Academia and the pursuit of intellectual knowledge were then very much the domain of the church, of people trained in divinity and theology. If this were the understanding of the great men of the day, the great thinkers, how dare anybody speak out

James Ussher, the Anglican church leader who in 1650 claimed to have proved that the Earth had been created in 4004 BC. Three centuries later, radiometric dating proved it was in fact billions of years old. SPL

against it? The date of the Earth's creation was even printed in the margins of the Authorised, or King James, Bible.

Now cut to the late 1700s. William Smith, an engineer and surveyor who is today regarded in Britain as the father of geology, observes that engineers and workers building canals keep cutting through the same sequence of rock formations. Each rock formation has its own distinctive set of fossils. It doesn't matter where Smith goes, he is able to recognise and match the formations. These cuts into the fabric of the land are beginning to show a remarkable, systematic pattern. It makes sense to name these distinctive sets of layered rock, and in 1822 the term Carboniferous is coined for one particular set of formations that have in common a recognisable variety of distinctive fossils. These rocks happen to be coal-bearing, hence the term 'carbon-iferous'.

Discovering the Age of the Earth

This would mark the beginning of a new branch of natural sciences called stratigraphy, the systematic study of the way in which layered, or stratified, rocks are organised in space and time. Many great men of science were to become preoccupied with this new kind of enquiry, making it a hallmark of the Victorian era. Long before the end of Queen Victoria's reign, all the great periods of Earth's geological history will have been discovered and named.

In the same year that the Carboniferous was named, the term Cretaceous, meaning chalk-bearing, was coined by Belgian geologist Jean-Baptiste-Julien d'Omalius d'Halloy after he had studied the strata of the Paris Basin. Chalk characterises the England–France–Belgium area around the English Channel and is very distinctive.

In 1829 the term Jurassic, after the Jura Mountains in France, was coined by French geologist and mineralogist Alexandre Brongniart, and in 1833 the labels Pliocene, Miocene and Eocene were introduced by the Scottish geologist Sir Charles Lyell. These were based on the relative abundance of identifiable shells within the rocks around Paris. Lyell noted that about 90 percent of fossilised shells in Pliocene rocks were of species that still existed, compared with only 60 percent in older Miocene rocks, and 30 percent in even older Eocene rocks. Clearly, as you went back in time there were fewer and fewer fossils that could be related to species still in existence. This was a most important realisation, and Lyell, a contemporary of Charles Darwin, would come to be acknowledged as the father of modern geology. His book *Principles of Geology*, the first volume of which was published in 1830, spawned and empowered the science.

Although Lyell never came to New Zealand he has a particular association with this country, having used data from the 1855 Wairarapa earthquake to establish beyond all reasonable doubt the causal link between earthquakes and faults, fractures in rock strata that show evidence of relative movement. The earthquake, which was greater than magnitude 8.2, caused a rupture on the Wairarapa Fault, which runs up the east side of the Orongorongo, Rimutaka and Tararua ranges near Wellington. It resulted in up to 6.5 metres of vertical offset and 17 metres of sideways slip over a distance of at least 156 kilometres. The basis of Lyell's scientific writings on the matter was his close questioning in London of three men who had experienced the quake firsthand and made excellent observations: Edward Roberts, an engineer with the Royal Engineers; Walter Mantell, an employee of the New Zealand Company; and Frederick Weld, a farmer who established the first sheep station in the South Island, at Flaxbourne, and later became, for a year, the country's prime minister.

The year after Lyell had named Pliocene, Miocene and Eocene, the Triassic period was named by Friedrich August von Alberti of Germany – 'Triassic' because the rock layers included three distinctive formations – Bunter, Muschelkalk and Keuper – which were widespread in Germany. In 1835 the Cambrian period was named by Adam Sedgwick, the first Professor of Geology at Cambridge University, after the ancient kingdom of Cambria, or Cumbria, in northwest England. Four years later, the Silurian period was named by British geologist Sir Roderick Impey Murchison, a contemporary of Sedgwick and Lyell, who took the name from the fierce Welsh borderland tribe, the Silures. Murchison would later become famous as director general of the Geological Survey of Great Britain.

Discovering the Age of the Earth

That same year, 1839, the Pleistocene was named by Sir Charles Lyell, after he realised there was a discrete period of time between the older Pliocene and the Recent (or modern day), and the Devonian period was jointly named by Sedgwick and Murchison following a controversy between them over whether or not a distinctive set of rocks in Devon were Silurian (below the Devonian) or Carboniferous (above).

In 1841 the Permian period was named by Murchison after the Perm Basin, just to the west of Russia's Ural Mountains, where there was a distinctive set of formations with fossils quite unlike those in the older Carboniferous rocks below and the younger Triassic rocks above. Murchison had been invited to Russia by Czar Nicholas I and, along with Edouard de Verneuil and Alexander von Keyserling, two other notable geologists who happened to be lesser European aristocrats, asked to produce a geological map of Russia. It was in the process of doing this that they discovered the Permian.

In 1854 the Oligocene was named by Heinrich Ernst Beyrich, Professor of Geology at Berlin University in Germany. An authority on fossil shells, Beyrich recognised a significant interval of rock between the older Eocene and the younger Miocene. The Palaeocene was named by another German scientist, Wilhelm Philipp Schimper, in 1874, after he recognised a distinct period between the end of the Cretaceous and the Eocene.

Finally, in 1879 English geologist Charles Lapworth gave the name Ordovician period to the layer of distinctive rocks in Wales that overlie, and are therefore younger than, the Cambrian but are older than Silurian.

So gradually between 1822 and 1879 the fundamental subdivisions of geological time were described and named. Collectively, these periods cover 542 million years of the Earth's history. At the time the geologists didn't know this: they were merely recognising sequences of sedimentary rocks with distinctive fossils, and it took a while for them to realise that stratified rocks represented the orderly progression of geological time. In a sense, modern earth science is all about revealing the record of our planet as it is written in the rocks. You could think of the periods as the chapters of a book, the layers or strata as the pages, and the fossils as words on the pages. Earth scientists have learnt to read this book incredibly well, thanks to the Victorian scientists.

HAVING ESTABLISHED THAT ROCKS are a record of time, the next logical questions are: How old are the successive layers? And how long did it take for them to form? In order to address these questions I would like to digress and talk about a famous palaeontologist called Gideon Mantell. In 1822 Mantell and his wife Mary Ann went to a quarry near Cuckfield in Sussex, where, it is claimed, Mary Ann Mantell found a fossil tooth. Mantell was a country doctor with an abiding interest in fossils and geology. From his broad knowledge of anatomy, he quickly determined that this tooth was unlike any other known mammal or reptile tooth. He also collected other teeth and bones from the same quarry, many of which he acquired from the quarry workers. Using his keen artistic skills, he was able to slowly develop an impression of what the animal whose teeth and bones they were might have looked like.

He formally described his fossils in 1825 and named the creature *Iguanodon* because the teeth most resembled those of an iguana lizard. Scientists now know that *Iguanodon* was the

Mary Ann Mantell and Gideon Mantell: the couple's discovery of a dinosaur fossil led to the realisation that animals now extinct had once roamed the Earth. SPL

first dinosaur fossil to be recognised as such, but at the time the word 'dinosaur' did not even exist. It was not until 1842 that it was coined and formally used by the famous Victorian anatomist Richard Owen, curator at the British Museum of Natural History.

Gideon Mantell knew the creature whose remains he had found was not mammal or lizard, and that its species therefore no longer existed. This was an immensely important realisation: in one stroke, he had discovered the concept of extinction. Mantell's discovery changed the world forever. For the first time, humanity appreciated that life forms that had existed in the past were no longer present.

There is an interesting connection between Gideon Mantell and New Zealand. Mantell's son Walter spent many years in the country in various capacities, but most notably with the New

The famous *Iguanodon* tooth in the collection of the Museum of New Zealand Te Papa Tongarewa, seen here with Gideon Mantell's original label. MUSEUM OF NEW ZEALAND TE PAPA TONGAREWA (F.003915/09-11)

Zealand Company. When his father died, Walter inherited his belongings, including the famous *Iguanodon* tooth. Because of his association with the geologist James Hector, the founding director of the Colonial Museum (later to become the National Museum), Walter Mantell deposited many of his father's collections in the museum. To this day the tooth, an international icon of monumental significance to the history and philosophy of science, the holy grail of dinosaur-mania, is at the national museum Te Papa where, from time to time, this amazing, small yet immensely precious object is put on display.

Having established that the *Iguanodon* tooth was from an extinct animal, the next questions concerned the age of the animal and when it had lived. Consequently, a great deal of effort

Discovering the Age of the Earth

went into thinking about how to determine the age of a fossil. There was no easy answer.

The mid 1800s was a dramatic time. The human race had stumbled on to this new concept of extinction, but how did extinction relate to life? To the Bible? There was no mention of extinction in the Bible. Furthermore, how did extinction relate to evolution, the new theory just introduced by Charles Darwin? The truth was that Darwin *needed* the concept of extinction: it was a necessary consequence, indeed a requirement, of his theory of evolution and the progression of life. The discovery of the fossil remains of a large terrestrial animal totally unknown anywhere on Earth at that time was the proof he needed. Yet Darwin was wracked – and not just by the physical pains from the chronic illness he had caught while on a voyage to South America in the *Beagle* as a young man, a condition which would plague him for the rest of his life. He was conscience-stricken by the significance of his own observations and the implications of his own thoughts. Like all educated men of the day, he was steeped in theological doctrine and biblical teaching. What he was now thinking and writing about was at total variance with the biblical explanation of life.

TO RECAP, AT THIS POINT in the story we have stratigraphy, the geological time periods, and we have extinction. The Victorians had almost wrapped it up. The next step was to develop a method for dating natural objects and materials preserved in the rock record. Around 1898 in Paris, the Polish-born scientist Marie Curie started down the path that would eventually solve the puzzle. She and her husband Pierre Curie were studying curious uranium-bearing rocks from one of the world's most famous silver mines, near modern-day Jáchymov (Joachimsthal of old) in Czechoslovakia, when they discovered radioactivity.

Four years later physicist Ernest Rutherford, working at McGill University in Montreal with a young research colleague named Frederick Soddy, published a paper in which the two men suggested, on the basis of their studies of radioactivity, that it should be possible to work out the age of minerals. They knew that uranium decayed to lead. They realised that all you had to do was figure out how much uranium was in the crystal of a uranium-bearing mineral, as well as how much lead, then, because the rate of decay of uranium to lead was known, it should be possible to calculate the age of the mineral – how long since it had crystallised.

Not only that, Rutherford figured out a way of measuring atoms that is still used today. The instrument is called a mass spectrometer. Think of it as a giant sheep race. You feed in your atoms, they go along the race, and at the end there's a gate where a person separates out the big atoms from the little atoms – the sheep from the lambs. This is exactly how the mass spectrometer works, except that the gate is really a magnet. The big sheep go out one door and the little sheep go out another, and you just count them. Once you know how much uranium there is (big sheep) and how much lead (little sheep), you can figure out the age. This is the simple basis of radiometric dating.

Rutherford tried out a number of experimental dates on minerals and clearly demonstrated that his and Soddy's original prediction had been right, thus establishing the science of geochronology: the radiometric dating of minerals. Almost any mineral with a radioactive element can be dated, and the most common radioactive elements are uranium and potassium. The most widespread uranium-bearing minerals are zircon and monazite, both of which form in granites. The most common potassium-bearing minerals are potassium feldspar, micas (biotite and muscovite) and hornblende amphiboles. These minerals are remarkably abundant. Potassium is one of the eight elements that make up 90 percent of the Earth's crust. Potassium decays to argon, hence potassium-argon dating as opposed to uranium-lead dating. Nowadays there are numerous dating laboratories worldwide, and millions of radiometric dates have been made on almost as many different mineral samples.

Clearly, the dating of minerals and rocks is all about measurement of radioactive decay. But what is radioactive decay? Why does it happen? Enter Albert Einstein, another Victorian. It is not possible to talk about atomic theory without invoking Einstein. In 1905, the 26-year-old physicist was exploring ideas relating to light. What would it be like if you, an observer, approached the speed of light? That thought stream, strange as it was, led him to his famous equation $E=mc^2$. He established a direct relationship between energy and mass, and in one stroke explained how the sun works. Atoms in the sun are being transformed, and in so doing are releasing vast amounts of energy. Einstein worked out that the sun is a furnace in which four atoms of hydrogen are fused and turned into one atom of helium. In the process the mass (of four hydrogen atoms) is reduced by 0.7 percent, and this becomes energy. Hence there is a direct relationship between mass and energy. Natural radiation is produced either by atomic fusion, as in the sun, or by atomic fission, as in the spontaneous decay of uranium and potassium.

Scientists such as Marie Curie and Ernest Rutherford had been worrying and puzzling about how a rock or pebble, such as the samples from Jáchymov in Czechoslovakia, could contain radioactive elements. How could a solid entity be radiating energy? Surely something had to be happening to the mass: it had to be changing, reducing? Einstein established that this was indeed the case. A tiny little bit of mass was being sucked away as radiation. However, because the energy in the nucleus was huge, the energy being released was colossal.

There was now a sound method for dating minerals, but governmental and military interests more or less took over the new field of nuclear physics, and for almost 50 years there was, amazingly, no development or use of the knowledge for scientific purposes. Geophysics is primarily concerned with four physical characteristics of the Earth's crust – magnetism, electrical properties, seismology (wave physics) and gravity – and development of these key areas of inquiry focused on the potential firing of ballistic missiles from one part of the world to another. Military imperatives also drove the development of nuclear energy. Einstein was bitterly upset and distressed by what might happen if nuclear energy got into the wrong hands.

Mass spectrometers and technology for dating minerals and rocks did not, therefore, become common until the 1950s. But then dating really took off. Today there are dating labs all over the world, and many of my scientific colleagues are involved in dating rocks or processes or events. This information enables us to establish the genealogy of rocks – the whakapapa and history of the Earth.

By determining when events happened and how fast, we can gain vital knowledge about natural hazards in our environ-

Discovering the Age of the Earth

ment. We need dates in order to determine, for instance, how often the Wellington Fault might break as a consequence of earthquake activity. We have knowledge of only two events but through dating we can say with some certainty that the Wellington Fault moves about every 500 to 700 years. When you have only two events the results are not very statistically sound, so we have to be cautious in our predictions. Nevertheless, we can express things in terms of probability, and get a feel for how often this fault moves.

A much more robust discussion can be had on how often Taupo – the volcano – erupts. Lake Taupo is a magnificent body of water but it happens to occupy a huge caldera. In fact, it is believed to be one of the most dangerous volcanoes on Earth. Through dating we know Taupo has erupted 28 times within the last 26,500 years. On that basis we can say that it erupts, on average, every 900 years. That's alarming because the last time it erupted was 1800 years ago. But we also know from the dating work we have done that the largest gap between Taupo's eruptions has been 5000 years. Remember those figures though: the average is 900 years, it last erupted 1800 years ago, and that's based on events we know about. There may have been other eruptions for which there is no obvious record.

These are just two examples that illustrate the power of dating, and indeed the demand for dating of natural events and processes. I now want to touch on another example of how relevant dating is to our understanding of New Zealand's geological history. For almost a decade I have been working with Chris Adams, New Zealand's foremost geochronologist, exploring the origins of New Zealand's oldest sedimentary rocks, including the greywacke. Greywacke forms the axial ranges of both the South and North Islands, the Southern Alps, most of Otago, Canterbury, Marlborough and Nelson and, in the North Island, the Rimutaka, Tararua, Ruahine, Kaimanawa, Kaweka, Urewera and Hunua ranges. In fact, greywacke is the most common hard 'basement' rock in New Zealand and represents about 60 percent of our landmass. The schist rocks of Westland and Otago are just metamorphosed variants of greywacke.

So what is greywacke? Well, it is a sedimentary rock. This means that it was once sediment produced by the erosion of some pre-existing rock, a mother rock. The original greywacke sediment accumulated as sand, silt, mud and gravel on the sea floor, dumped there by ancient rivers draining an ancient continent, in this case Gondwanaland. It is named greywacke from an old German mining term 'grauwacke', meaning grey sandstone. As it was to German miners in centuries past, it is a boring, monotonous rock that nobody wants.

The interesting thing about the greywacke is that there is no source rock within New Zealand from which this vast flood of sediment can have been derived. So where did it all come from? For some decades geologists have been searching in Antarctica and Australia, and it now seems that the most likely answer is northeast Queensland. How can we possibly know this? From dating.

Here's how it works. We systematically sample the greywacke for careful analysis. This involves going out in what we geologists call 'the field', and using a geological hammer to collect small samples of rock, about half a kilogram in weight, from an outcrop. We number the sample carefully and record precisely where we collected it. Back in the laboratory we break down the sample using a crushing mill (not unlike a heavy-duty coffee grinder), and extract the minerals that contain either uranium or potassium, the common radiogenic elements used for dating rocks. The minerals are then dated using laboratories housed in universities in Australia. We call this kind of research 'provenance studies using detrital mineral age dating techniques' – 'detrital' because the minerals that are dated formed part of the sediment detritus the rock is made of, and are not minerals that have formed in the rock.

As an aside, the oldest rocks on mainland New Zealand, which are found in northwest Nelson, are about 508 million years old. They hail from the Middle Cambrian epoch and are trilobite-bearing – that is, they contain marine fossils of early Palaeozoic age. However Chris Adams has recently established that there are older rocks still – schists in Garden Cove on Campbell Island, which, at 700 kilometres south of Bluff, is the southernmost of New Zealand's subantarctic islands.

Another of our projects has been trying to find a way of characterising New Zealand nephrite, or pounamu. How do you distinguish this New Zealand nephrite from nephrite elsewhere, such as in China or Canada? We realised that a very simple solution would be on the basis of age. Because the Southern Alps, where the pounamu is found, are rising so rapidly – 10 to 15 millimetres a year – the rocks that comprise them must be very young. The reason for thinking this is that they are coming straight up and being eroded off. (It's a bit like somebody standing on a tube of toothpaste.) If you extrapolate backwards at that rate, and you don't have to go further than about four million years, you realise that everything would then have been very deep and very hot. Remember that as you go into the Earth, temperature increases about 20 to 25 degrees per kilometre.

Hence, if you could date New Zealand pounamu you could distinguish it from all other nephrite. Its cooling age ought to be very, very young.

In another project, with seven other scientists and three research students I am exploring the age of the land surface of the Chatham Islands, which lie some 850 kilometres due east of Christchurch. There are good geological reasons for thinking that the Chathams have been literally pushed out of the water within the last four million years. If we can prove it, the biological implications will be considerable. If we could demonstrate that there was no land there four million years ago, we would gain all sorts of information about rates of dispersal from the mainland and elsewhere out to the Chathams – as well as rates of both colonisation and evolution.

After our second field season we have had a breakthrough, discovering a new formation on the southwest corner of the biggest island in the group, Chatham Island. This formation is more than 200 metres above sea level yet less than a million years old. It is a marine sediment, a limestone, and rich in fossils with modern-day species of scallop and oysters, so we will be able to determine quite precisely how old it is. From this we should learn a great deal about how fast the Chathams have been pushed up out of the water. As well, we are using molecular biology on selected plants, insects and forest birds to try and establish genetic distance between native species in the Chathams and comparable species on the New Zealand mainland. That, too, will be a measure of age.

WE HAVE COME A LONG WAY from the ideas about the age of the Earth that developed in the 1600s. Our ability to understand life has been greatly modified by our understanding of the history of the planet itself. Unravelling the rock record with the advent of, first, stratigraphy and then geochronology has totally transformed our comprehension of the natural world. We now know through radiometric dating that the Earth is 4.53 billion years old. Despite all attempts, nothing older than this has been found in our solar system. We also know that Gideon Mantell's *Iguanodon* is 125 million years old (Cretaceous), and that the dinosaurs became extinct 65 million years ago, at the very end of the Cretaceous period. The major periods of geological time from the Cambrian to the Recent are now fully calibrated, just like our calendar: the Cambrian commenced 542 million years ago, the Ordovician 490 million years ago, and so on. We also know that most of the abrupt changes in fossils from one period to the next, such as Permian to Triassic, were due to global extinction events, and that these in turn can be attributed to terrible catastrophic accidents of nature, such as impacts from meteorites and comets, or vast volcanic outpourings of basalt.

Perhaps the biggest breakthrough in our post-Victorian understanding of how the Earth works was the advent of plate tectonic theory in the mid 1960s, and even this was built on dating. In the process, an unusual but nevertheless very simple question arose. It too was to do with the age of the Earth, but it was more subtle and more cryptic, concerning something we cannot easily reach because it is buried by water and sediment. The question was this: why is the ocean floor no older than 180 million years? Comprehensive exploration had established two quite startling but related facts. First, the ocean floor is entirely comprised of one type of rock, namely basalt. Scrape or poke below the obvious sand, silt and mud on the sea floor and what is always found is basalt. Secondly, nowhere is the basalt older than about 180 million years. This needed some explanation.

The answer turned out to be simple. The ocean floor is behaving like a giant conveyor belt, rising as fresh lava along the mid-ocean ridges, and being destroyed by subduction – literally sucked back into the earth – along deep ocean trenches. This moving process is called sea-floor spreading.

Plate tectonic theory explained the motion of the entire Earth's crust. At the time the theory was first enunciated it was seen as bold, but it offered an extremely satisfying, holistic, all-embracing explanation. Since then, modern technology using satellites and lasers has established beyond all doubt the directions in which all places on the surface of the Earth are moving, and the rates at which they are moving. Around New Zealand, the Earth's crust is moving at four to five centimetres per year, about the rate at which our hair and fingernails grow. This is not very fast, but imagine if you have a million years to play with, or 180 million.

Plate tectonics has revolutionised comprehension of the natural world in our lifetime but we can thank the Victorians for establishing both the theoretical and practical tools for cracking it. Timing and age are everything!

Hamish Campbell *is a geologist and research scientist with GNS Science in Wellington. Born in Christchurch and raised in Dunedin, he was educated at Otago, Auckland and Cambridge universities, specialising in palaeontology. Much of his research concerns the older rocks and fossils of New Zealand, those that are between 300 and 150 million years old. Another research interest is the geology of the Chatham Islands. However, he is best known as the geologist at the Museum of New Zealand Te Papa Tongarewa, where he has become a science communicator, helping promote geology and provide geological information to the public.*

Einstein and the Eternal Railway Carriage
—*Lesley Hall & Richard Hall*

LESLEY HALL:

WHEN ALBERT EINSTEIN DIED in 1955, a newspaper cartoon showed the Earth with a sign on it which said simply: 'Albert Einstein lived here.' Forty-five years later, as the new millennium dawned, Einstein had become iconic, the poster boy for science, acclaimed by *Time* magazine as the most important person of the twentieth century, 'the genius among geniuses who discovered, merely by thinking about it, that the universe was not as it seemed'.

My interest is not just in Einstein's contribution to physics but the social and political environment in which he studied and worked. British sociologist Liz Stanley argues that a biography worthy of the name should firmly grasp the cup of plenty that is a person's life. All aspects should be considered, including things that may at first glance seem incidental. Is, for example, the fact that Einstein was Jewish irrelevant to the story of his life? He was born in 1879, the year the term anti-Semitism was first coined by German agitator and writer Wilhelm Marr, and some biographers claim that his Jewishness was the major cause of his long wait for the Nobel Prize. After being rejected eight times, he finally received the award in 1922 for his contribution to theoretical physics – not for the general theory of relativity or the special theory of relativity, as you would imagine, but for his discovery of the law of the photoelectric effect. But more of that later.

Biographies are perennially popular (a result, perhaps, of the gossip factor) but readers should bear in mind that the picture of the person that emerges depends on the interests of the biographer, his or her relationship to the subject, and the range of source material available. Hundreds of books have been written about Einstein and some, especially the earlier ones, present sanitised versions. Einstein's executors managed to prevent the publication of a book about him by his elder son Hans Albert, presumably because it tarnished the saintly picture they were keen to promote.

Little mention is made in early biographies about the relationship between Einstein, his two wives – he was living with the second before divorcing the first – his sons, and his stepdaughters. Another glaring omission until recently was the birth of a daughter out of wedlock to Einstein and his fellow student Mileva Marić, who later became his first wife. This information became available only when Marić's letters entered the public domain in the 1980s. A different picture of Einstein may emerge once again when the letters between him and his elder son, and between him and his second wife Elsa, also become available for public scrutiny.

Does exposing negative aspects of Einstein's character minimise the significance of his contribution to science and to humanity? I don't believe so. Biographer Denis Brian (*Einstein: A Life*), like me, favours a 'warts and all' approach, claiming that revealing Einstein as a more compelling, complicated and controversial individual than previously thought – still in all his glory but with his halo slightly askew – makes him an even more endearing subject.

Einstein and the Eternal Railway Carriage

I could tell Einstein's story by focusing on his different roles – Einstein the socialist, the pacifist, the Zionist, the father, the husband, the friend. Each of these would reveal particular facets of his personality. But I must be brief, so a little of his family and educational background will have to suffice – a brief summary of some of the circumstances that led up to 1905, *annus mirabilis*, the year of miracles, in which he published five scientific papers while working at the patent office in Bern, Switzerland. So if you're sitting comfortably, let us begin.

Albert Einstein was born in Germany on 14 March 1879. He was a late developer, not talking until the age of two or three, depending on which biography you read and on Einstein's own version of events. His parents were initially concerned that, because of his lateness in talking, he was perhaps a little slow-thinking. (Presumably they later changed their minds.) Einstein attended a Catholic primary school, where he was not happy. His later reflection on education gives a clue to the source of his discontent:

> *To me, the worst thing seems to be for a school principally to work with methods of fear, force and artificial authority. Such treatment destroys the sound sentiments, the sincerity and the self-confidence of the pupil. It produces the submissive subject.*

The young Einstein was not thought brilliant by his teachers, and his headmaster apparently said that he would never make a success of anything. The reason for his unhappiness, and his teachers' poor assessment, was that Einstein was unorthodox in both his thinking and in the way he worked. He appears to have

Albert Einstein aged 14. He was not thought brilliant by his teachers but educated himself in science with the help of relatives and family friends. SPL

been a critical thinker and to have challenged authority from a very early age, rarely taking a teacher's word for something without question.

His parents appear to have been quite liberal in their approach to child-rearing, encouraging him to think independently and allowing him to develop his own interests. At the age of five he was given a compass by his father, and he would later credit this with developing his interest in the forces of nature. Einstein's German high school was strict and very formal. Reportedly the teachers were heavy-handed and impatient, and he was considered by most of them to be disrespectful.

His Greek master, in an echo of his elementary school principal, said he would never amount to anything. From the time he was 12, science and philosophy became part of Einstein's life. Some biographers argue that he was largely self-educated in science, and this was possible because of support from his family and their friends. An uncle introduced him to algebra, and a medical student, Max Talmey, a weekly dinner guest of the Einstein family, to popular science books.

At the age of 16 he finally dropped out of school and went to join his parents, who were by then living in Italy. After being turned down by the Swiss Federal Polytechnic (Eidgenössische Polytechnische Schule) in Zürich, he attended a secondary school in Aarau, Switzerland, an institution he found much less rigid than its German counterparts and consequently more to his liking. The headmaster, Professor Jost Winteler, with whom Einstein boarded, was a freethinker and liberal in his approach to learning. (Einstein's sister Maja would later marry Winteler's son Paul.)

Einstein appears to have been liked by both pupils and teachers, and after gaining his diploma he decided on a teaching course at the Polytechnic. There his lecturers were extremely irritated by Einstein's failure to attend lectures and thought him arrogant, rebellious and outspoken. One professor reportedly told him, 'You're a clever fellow, Einstein, but you have one fault – you won't let anyone tell you a thing.' Another lecturer described him as a lazy dog who never bothered with maths at all. Einstein did neglect maths, considering it overspecialisation, and reasoning that he didn't have time to learn everything. It was only later that he would realise his mistake. In the early years of their marriage, his wife Mileva Marić helped him with his mathematics. Later, he enlisted the assistance of friends

Einstein believed the Polytechnic lecturers were limited in their thinking and out of touch with the most recent scientific theories and research. Michael White and John Gribbin, in their biography *Einstein: A Life in Science*, suggest this so-called arrogance can be given a different, more positive interpretation, namely that Einstein knew best what worked for him and had confidence in his own approach. Einstein's method could, they argue, be described as showing maturity and self-awareness rather than arrogance or disrespect. Nevertheless, after passing his exams – cramming a friend's lecture notes because he hadn't been to the lectures – and getting a diploma, Einstein found it impossible to get an academic position. Several reasons have been put forward but the consensus seems to be that the rejection came about through a combination of anti-Semitism and his poor relations with teachers.

Einstein eventually got a job at the patent office in Bern in 1902. He married Marić in 1903 after the birth of their daughter Lieserl. What happened to Lieserl is the subject of speculation. One claim is that she was born mentally handicapped and died of scarlet fever at the age of two; another is that she was adopted. The couple later had two sons, Hans Albert and Eduard.

Marić's contribution to Einstein's scientific work is contested. Some biographers suggest she played a peripheral role, while others such as Andrea Gabor (*Einstein's Wife*) claim she subsumed her own scientific ambition and interests to those of her husband, and must have been exceptional in her own right to have got as far academically as she did. Einstein himself credited his wife with solving his mathematical problems, and she probably proofread his papers. It seems likely that she was also a sounding board for his ideas.

Gabor goes further, saying there is fragmentary evidence to suggest that the original versions of Einstein's famous articles on the photoelectric effect, Brownian motion and the theory of relativity were signed Einstein-Marity, Marity being the Hungarian version of Marić. In 1905, after the paper on special

Mileva Marić in 1896 aged 21, when she left her home in Zagreb for education in Switzerland. While a student at Zürich's Federal Polytechnic she met Einstein, and the couple married in 1903. They would separate acrimoniously in 1914 and divorce in 1919. SPL

relativity was completed, Marić is said to have boasted to her father that she and Einstein had just finished some important work that would make her husband world-famous.

Einstein eventually got a position at the University of Bern in 1908: this was the beginning of his academic career and closer involvement with the scientific research community. In 1909 he became Professor Extraordinary at Zürich, and in 1911 Professor of Theoretical Physics at Prague, returning to Zürich the following year to fill a similar post. In 1914 he was appointed Director of the Kaiser Wilhelm Physical Institute and a professor at the University of Berlin. In 1919 his marriage to Marić was dissolved and he quickly married his cousin Elsa Löwenthal. In 1933, dismayed by the rise of Hitler, the couple emigrated to America where Einstein became Professor of Theoretical Physics at the Institute for Advanced Study in Princeton, New Jersey.

Albert Einstein and his second wife Elsa, photographed in 1921, two years after their marriage. SPL

Einstein's attitude to science and academia appears to have been influenced both by the way he himself had been taught science and by the way he was treated by other scientists. There is evidence that there was a considerable amount of professional jealousy, possibly another reason for the delay in awarding him a Nobel Prize. It has also been argued that some of the scientists involved could not understand relativity, and opposed making an award for Einstein's paper in case he was later proved incorrect. White and Gribbin argue that Einstein was also passed over for the Royal Astronomical Society Medal in 1920 because of the significant influence of detractors, who objected on the basis of envy or political influence: anti-Semitic feeling was very strong in some quarters.

EINSTEIN DOES NOT SEEM to have viewed science as a superior belief system. Nor did he think scientists should be put on pedestals, arguing that non-scientists had as much right to discuss scientific issues as so-called experts. As for the teaching of science, he argued that lessons should be interesting. A young mind, he said, should be spared formulae altogether; demonstrating a 'pretty experiment' was the recommended method for capturing young minds.

Einstein's views of science and scientific culture are as relevant today as they were 100 years ago. He was concerned with what he saw as a corrupting influence on the scientist: the need to be successful. In May 1927, when there was debate as to who should succeed Max Planck as physics professor at the University of Berlin, he wrote to a friend:

I am not involved, thank God, and no longer need to take part in the competition of the big brains. Participation in it has always seemed to me to be an awful type of slavery, no less evil than the passion for money or power.

He would also have nothing to do with the production of what he called 'waste paper' in the form of academic publications, which he viewed as the great blemish of the academic world.

A number of biographers have discussed Einstein's genius, and Denis Brian has a chapter devoted solely to a discussion of his preserved brain. Underlying the discussions is the desire to explain how Einstein managed to achieve work of such sheer brilliance. Psychologist Anthony Storr has postulated that the theory of relativity could have emerged only from a personality with a strong sense of detachment – someone who could stand back and observe from the outside. Such behaviour, Storr says, is indicative of schizophrenic tendencies. It has also been suggested that Einstein may have had Asperger's Syndrome, a mild form of autism. Einstein himself, when asked to account for his discoveries, said that he achieved them through both intuition and inspiration:

> *Sometimes I feel I am right but I do not know it. I'm enough of an artist to draw freely on my imagination, which I think is more important than knowledge. Knowledge is limited; imagination circles the world.*

RICHARD HALL:

THINKING OUTSIDE THE SQUARE is absolutely essential to understanding relativity, and that's where genius comes in. When we think of relativity, we tend of think immediately of Einstein, but it was not Einstein who first introduced relativity to the scientific community, but Galileo in the early seventeenth century. One day, while working for the Venetian navy, Galileo began to make a few observations about things normally taken for granted. He noticed, for example, that at night when he was in his cabin and the ship was sailing but the sea was very calm, it was impossible to tell the ship was moving without looking outside. You may have experienced this yourself when flying in an aircraft. Galileo termed it the principle of relativity. It meant there was no such thing as absolute rest: all motion measured in the universe was relative. We cannot take any point and say, 'This is it, this point isn't moving and everything else is.'

A long time later, towards the end of the nineteenth century, physics was beginning to get into a few problems to do with the speed of light. Scientists had managed to measure the speed, and today we know the figure is close to 300,000 kilometres per second. But then when a few other experiments were carried out they began to discover some rather unsettling things about light. For example, if you were approaching a light source, you would expect the light to intercept you at a speed greater than 300,000 kilometres per second, in much the same way as two cars both travelling at 50 kilometres per hour heading towards each other will intercept at 100 kilometres per hour. Similarly, if you imagine yourself moving away from a light source, you would expect the light to overtake you at a speed of *less* than 300,000 kilometres per second.

However, it didn't matter in which direction the scientists moved, whether towards the light or away from it, the light moved towards or away from them at 300,000 kilometres per second. This seemed to be a really unusual property, and it began to produce some things that were difficult to explain. From this, we come to what we might call 'The Light Paradox'. I will use some of the ideas Einstein himself used to try and explain this.

Imagine you are looking at your face in a mirror. The light bounces from your face on to the mirror and then back into your eyes. That sounds easy enough to understand, but now imagine yourself in a spacecraft accelerated to the speed of light. If the speed of light is fixed at 300,000 kilometres per second and you are travelling at 300,000 kilometres per second, this means the light leaving your face is moving at the same speed as you. The light can never reach the mirror, and that of course is a violation of the principle of relativity: when you reached the speed of light your image would disappear from the mirror and you would know how fast you were moving without looking outside. If you are moving along at the speed of light and light can't catch up to the mirror, it means that the speed of light – as far as the person who is travelling is concerned – is zero: it's not moving anywhere. But experiments have already told us that the speed of light is absolutely independent of whether you are moving or not. So herein lies a paradox.

Let me introduce you to another paradox that arises from this. Einstein, when trying to describe it, talked about trains and thought about railway carriages. Imagine a railway carriage. In this railway carriage there is a light. The light source is positioned at the centre of the carriage, and at each end of the carriage there is a door. When you switch on the light source, the light travels along, hits the light sensor at the door, and the door opens. Now, because the light is positioned right at the centre of the carriage and light moves at a constant speed, it should hit both detectors at either end of the carriage at exactly the same time, and therefore both doors should open together.

But what happens if the train is moving? When the light leaves the source, it is independent of the source and moving through space at a constant speed. To an outside observer, the back end of the carriage is coming towards the light while the far end is moving away, therefore the light will reach the back end of the carriage before the front, and the back door will open first. But this would violate the principle of relativity because a person inside the carriage would know that they were moving without looking outside. The person inside the carriage would say, 'This is silly. The light is exactly the same distance from either end. Why, when I switch on the light, does the back door open first?' For the passenger, both doors should open at the same time. But to an outside observer the light would reach the back end of the carriage first and the back door should open first.

So what is the answer? Do both doors open at the same time, or does the back door open first? I'm not going to tell you the answer yet – I'm going to let you think about it. First, let's look to see how Einstein began to work out this paradox. He started off with some basic assumptions upon which the whole of relativity is based.

Assumption 1: No matter how light propagates when you are standing still, it propagates exactly the same way when you are moving. (This is the principle of relativity.)

Assumption 2: Light is always propagated in an empty space with a definite velocity, namely 300,000 kilometres per second, which is symbolised in Einstein's famous equation as 'c'. This speed is independent of the state of motion of the emitting or receiving body.

Einstein realised that in order to answer the paradox he had to explain how the man in the spacecraft, moving at the speed of light, would see light move away from his face towards the mirror at the speed of light, while at the same time an observer on the ground would see the light leaving the face of the astronaut at exactly the same speed? Let's think about the speed of light. Speed is two quantities, distance divided by time, as in kilometres per second. Einstein realised that if speed were to be the same for all persons, then distance and time had to be different. Perhaps the traveller observed different times to the person on the ground? When Einstein began to think along these lines, he was immediately challenging the whole of physics before him, because in classical mechanics space and time intervals were absolute and unchanging and, if anything, the speed of light was relative. Einstein said the opposite: the speed of light was absolute, and all the other quantities relative.

To understand what he was saying, let's think about a metre rule. If asked, you would probably say that a metre rule where you live would be exactly the same length as a metre rule in any other part of the country, or the world. But that's an assumption on your part. How do you know? If we were in a room together and, by magic, I could decrease the size of everything by 50 percent, you wouldn't notice any difference because everything around you would have shrunk by the same amount. The only way you would realise that everything around you had been shrunk by 50 percent would be by looking outside, and then you would see that everything else looked rather large.

The same thing applies to time. If I were to slow time by 50 percent, you wouldn't notice because the clocks would be moving at half-speed, your heart would be moving at half-pace, your brain would be moving at half-pace, even the atoms in your body would be moving at half the pace: everything would appear absolutely normal. The only way you would notice that time was moving at a different rate would be to look outside and see everything else apparently moving at super-speed, like a film that had been cranked up. So what Einstein was arguing was that these fundamental things, like the length of a body or the rate at which time moves, things we think of as constant everywhere, are not necessarily so.

Following this line of thinking, Einstein developed the special theory of relativity. Let's see how the special theory of relativity explains the light paradox. Let's first of all imagine the traveller. He is holding the mirror in front of his face, which may be only half a metre away, and he is sitting inside his spacecraft.

How fast he is moving relative to everything else doesn't matter: the point is that the mirror is only half a metre away from him, and the light leaves his face and travels to the mirror almost instantly because of the short distance.

The person on the ground, however, sees things entirely differently. Because light moves at a finite speed, the mirror is in a sense moving through space and time. So by the time the light reaches the mirror, it has had to travel a much greater distance – for example, in one second the mirror will have moved 300,000 kilometres. Hence, the stationary observer on the ground sees the light travelling along a longer path and observes more time elapsed. This is why Einstein said that moving clocks run slower than stationary clocks.

Let's now return to the train – the light and the two doors. Do both doors open at the same time? Or does the back door open first? The answer is that both are correct. For the person on the train, both doors open at exactly the same time, but for the person standing outside, the back door opens first. You see, none of these things – the light being switched on, the sensors opening the doors – is a single event. They are all separate events. There are no instantaneous interactions in nature, and simultaneous events in one frame of reference may not be simultaneous when viewed from a different frame of reference. Einstein himself said that relativity teaches us the connection between the different descriptions of one and the same reality.

Let's see how this dilation of time works. Einstein provided an equation into which you could put the numbers and work out how much time slows down if you are travelling at high speeds. The star Alpha Centauri is, at 4.3 light-years away, the nearest star to our solar system. When we talk about a light-year, we are talking about the distance light travels in one year, and because light moves at a constant speed in a vacuum – 300,000 kilometres per second – to work out how far it travels in one year, you merely need to work out the number of seconds in a year and multiply that by 300,000. The answer, incidentally, is 9467 billion kilometres. A light-year, therefore, is a measurement of space, and Alpha Centauri is 4.3 light-years away. It is also a measurement of time. Because light from Alpha Centauri takes 4.3 years to reach us, we see it as it was 4.3 years ago.

Now the effects of special relativity are there all the time, but they don't really become significant until we begin to move at an appreciable percentage of the speed of light. Let's start off by looking at someone moving at 50 percent of the speed of light. For the person staying at home, the observed journey time to Alpha Centauri would be 8.6 years, but for the traveller the journey would take only 7.5 years. Now, let's increase our speed to 90 percent of the speed of light. The journey now, as far as the person observing back home on Earth is concerned, would take 4.8 years, but for the traveller only 2.1 years would elapse.

And the closer we get to the speed of light, the greater these effects become. Let's take a speed of 99 percent of the speed of light. For the person staying at home, the journey time is almost equivalent to the speed of light – 4.34 years – but for our traveller it is only just over 7 months. At 99.9 percent the speed of light, the journey time is reduced to 22 days, and at 99.9999 percent to 16.8 hours.

Let's think about this more closely. The person staying at home – or we, the people on Earth – will see the spaceship leave Earth and accelerate to almost the speed of light, and we will measure the journey to Alpha Centauri as taking 4.3 years. But for the person travelling on board the spaceship, the journey will take less than a day – in fact, they could make a round trip without taking food with them. They could theoretically go to Alpha Centauri, have a cup of tea, turn around, and in just under 34 hours be back on Earth again, and they will have aged just over a day. But when they return, they will find that almost nine years have elapsed on Earth. This is what time dilation does for us.

Now I want you to think a little further along these lines. I just mentioned that if we travelled at 90 percent of the speed of light, the journey for the traveller would be only 2.1 years. But this doesn't make sense because we have just said that Alpha Centauri is 4.3 light-years away. Surely for the traveller to get there in two years, he or she must be travelling at twice the speed of light? But no, they look at their clocks and they are moving at only 90 percent of the speed of light. The difference is this: when they accelerate to 90 percent the speed of light, they discover that Alpha Centauri is not 4.3 light-years away at all. It is only 2.1 light-years away. Remember the length of a metre rule? The length – the actual space itself – appears to be different for the traveller.

So concepts of space and time depend upon your motion within space. In much the same way as electricity and magnetism had been found to be two manifestations of a single force, electromagnetism, Einstein had discovered that space and time were two manifestations of the same entity which he called 'space-time'. One of the consequences of Einstein's special theory is that you cannot reach the speed of light. Let's see why this is so. The observer on Earth sees time on the spacecraft slowing down, and the faster the craft goes the more time slows down. Using Einstein's formula it can be calculated that when the spacecraft reaches the speed of light, time would have halted:

the last second lasts forever. In other words, if you could actually reach the speed of light, time would literally stop. You would be frozen in time. That's what the person on Earth would observe. But of course, the traveller would see things entirely differently. Time would appear to move at the normal rate, but as the spacecraft went faster the thrust from the engines would appear to decrease. Imagine you are pushing a trolley at a supermarket. You are applying a constant force but the trolley appears to be getting heavier and heavier. You could deduce that someone must have put more items into the trolley.

Well, it works this way: as you approach the speed of light and apply more thrust to your engines, instead of going faster and faster the spacecraft becomes heavier and heavier. Its mass increases. An object accelerated to the speed of light would have an infinite mass. And, of course, to accelerate an infinite mass to the speed of light would require an infinite amount of energy.

This isn't just theory. It is the sort of thing scientists actually do these days in huge accelerators. Tiny particles can be accelerated until they are moving close to the speed of light, and when they get to the collector plate at the other end they thump into it, indicating an enormous mass increase. When this is measured, it is found that the mass in the particle has increased by precisely the amount Einstein calculated. What is happening is that more energy is being put in – and what we are ending up with is mass.

This is what led Einstein to formulate the most famous equation of the twentieth century: $E=mc^2$. In other words, matter and energy are manifestations of the same thing. You can convert energy into matter, and matter into energy. And so with the spacecraft, as we're trying to go faster what we're actu-

In 1911 an international physics conference in Brussels was attended by an array of ground-breaking scientists, including Albert Einstein (second from left, standing), Marie Curie (second from left, seated) and behind her Ernest Rutherford. It was sponsored by Belgian chemist Ernest Solvay. PROF. PETER FOWLER/SPL

ally doing is turning energy into matter. This was a revelation because for the first time astronomers could understand what powers the stars. Essentially stars are gigantic thermonuclear engines converting matter into energy. Prior to this, try as we would we couldn't work out how stars could maintain such an enormous energy outflow. Our sun is a typical star, and every second it converts something like 4 million tonnes of matter into energy. Think about that – every second the sun becomes 4 million tonnes lighter, and it has been doing that for billions of years. But then the sun's a pretty big object.

SO THE STORY SO FAR: Einstein takes space and time and joins them together as a single entity which he calls space-time, and then he joins matter and energy together in the equation E=mc^2, where c is the speed of light. This brings us to gravity, because relativity challenges the very foundations of Isaac Newton's notion of gravity. Where Newton saw gravity as a force interacting between two masses, Einstein's general theory of relativity describes gravity as the geometry of space-time.

To help you understand the general theory of relativity, I would first like to tell you a story about an imaginary place called Flatland. Flatland is an island city in a universe in which there are only two dimensions. The city has length and breadth but no height. All the structures and creatures that live there are completely flat. They have shape, such as circular or square, but they do not exist in the third dimension. Indeed, because they can only ever experience two dimensions, the Flatlanders can't even imagine what a three-dimensional object looks like.

The scientists at Flatland want to know whether their universe is finite or infinite. They decide to embark on an expedition to see if it has an edge or boundary. Two Flatlanders are sent on the expedition, each in a separate ship (which is also perfectly flat). They travel away from Flatland City along paths that are perfectly straight and parallel to each other.

After a while the travellers notice something very curious: the distance between them is decreasing. Although they are travelling in perfectly straight lines parallel to each other, somehow they are also moving towards each other. Or perhaps it is the space between them that is vanishing? Eventually the two ships cross paths and then begin to slowly move apart. Again they check their instruments, and according to the readings they have not deviated from moving in straight lines along paths that were originally parallel to each other.

The paths they now travel along are moving them apart. However, despite the fact that both are travelling in straight lines, this outward drift eventually halts and they begin to move back towards each other again. Again, they cross paths and begin to move apart once more. Then one day, the thoroughly confused travellers see something amazing. Directly ahead is an island city. It is their city – Flatland.

Have you worked out why these strange things have been happening? Although the Flatlanders exist in only two dimensions their universe is built in three. Their realm of existence is like a skin on the surface of a sphere. Flatlanders cannot imagine what a sphere looks like but they can calculate its geometry. If you travel along a straight line on the surface of a sphere you eventually arrive back where you started. Great circles drawn around a sphere can start out parallel to each other but, like the lines of longitude on Earth, will bisect each other at two points.

The Flatlanders have their answer: their universe is finite but it is also unbounded – it has no edge. And because it has no edge, no matter where a Flatlander is he or she will always appear to be at the centre: the centre of the Flatland universe is everywhere. If Flatlanders were to travel in a straight line for ever and ever they would never be able to leave their universe. To us (three-dimensional creatures) the Flatlanders would simply be running around in circles.

We face the same problems as the Flatlanders in understanding our universe. Einstein's space-time is constructed in four dimensions (and there may be more). Because the Flatlanders existed in only two dimensions they couldn't visualise a three-dimensional object. In a like manner we find it difficult to visualise something constructed in four or more dimensions.

The best way I can explain how our universe is constructed is to reduce it to two dimensions. Imagine a tight piece of rubber such as a trampoline, and imagine this trampoline represents space. If I roll a ball across the trampoline it will travel in a perfectly straight line. Now let's imagine that I put a great big lead cannonball at the centre of the trampoline. This will cause a dip in the trampoline. If I have another ball and put it on the edge, it will roll towards the cannonball.

We can, of course, see this is happening because the surface of the trampoline is now curved. But let's assume we can't see the trampoline. When the ball moves towards the cannonball we might say, 'Hey, there seems to be a force of attraction between the cannonball and our little ball.'

Now let's imagine that instead of putting the ball on the edge, we flick it around the cannonball. On a trampoline, the ball will gradually spiral into the hole where the big cannonball is. But in space there is no friction, so the ball will simply rotate around and around it; we would say it's 'in orbit'. So Einstein imagines space and time as being curved around massive objects – the Earth orbits around the sun not because there is a force of attraction between the sun and the Earth, but because the sun is a massive object and space is curved around it: Earth is moving essentially in a straight line in curved space-time. It's moving in what we call a geodesic – like the guy on the motorbike on the 'wall of death' we used to see in sideshows. Einstein said of space-time, 'When a blind beetle crawls over the surface of a globe, he doesn't realise that the track he has covered is curved; I was lucky enough to have spotted it.'

So what does this tell us about our universe? Most physicists today think that our universe is similar to Flatland in that it is finite but unbounded. No matter where you are in the universe, you will appear to be at the centre: the universe has no edge. And if you were to travel in a straight line in space, according to the general theory of relativity you will eventually come back to where you started from. (When the general theory of relativity was first proposed, astronomers using the most powerful telescopes in the world tried to observe our home galaxy by looking out into the depths of space. The theory was that if you had a big enough telescope, you should be able to see the back of your own head.) In reality, because the universe as we know it appears to be finite in age, this would not be possible.

This general theory of relativity is what current models of the universe are based on. While a lot may appear nonsensical and beyond common sense, the Hubble space telescope and others are using the general theory of relativity to make new discoveries about our universe. We are now seeing things in space which we could only see if the theory were correct. It was recently announced, for example, that a team which included two amateur astronomers in Auckland had discovered a planet around another distant star 15,000 light-years away. They achieved this by using the effects predicted by the theory of general relativity – the curvature of space-time.

So in summary, Einstein revolutionised the way physicists view the universe in which we live. He created, in the words of White and Gribbin, 'an entirely new way of visualising reality, backed up by rigorous mathematic formalism – an achievement sometimes referred to as the greatest intellectual effort of any single human brain.'

Lesley Hall & Richard Hall

Lesley Hall *is acting programme director of Gender and Women's Studies at the School of Education Studies, Victoria University of Wellington. Her teaching and research interests include autobiography, biography and oral history, and feminist analyses of science, medicine and technology. She has recorded oral histories of New Zealand women scientists, and is coordinator of a project to record oral histories of Emeritus Members of the Royal Society of New Zealand. She is president of the National Oral History Association of New Zealand, a founder member of the Phoenix Astronomical Society and a key player in the construction of Stonehenge Aotearoa.*

Richard Hall *is an astronomer and communicator with a gift for making complex scientific information understandable and enjoyable for the layperson. He is the author of* How to Gaze at the Southern Stars, *which was an instant success and is now into its third printing. Hall is president of the Phoenix Astronomical Society, a member of the Astronomy Standing Committee of the Royal Society of New Zealand, and a Royal Society Science Communicator. He is currently involved in developing Stonehenge Aotearoa in the Wairarapa as a centre of discovery and learning – from ancient star lore to the mysteries of the cosmos.*

Schrödinger's Cat
—Tom Barnes

BEFORE I GET ON TO THE SUBJECT of quantum mechanics, I would like to mention some general matters to set the scene. First, complacency in thinking we understand nature is invariably jolted by nature taking scientific discovery along paths previously both unheard of and unimaginable. This is perfectly illustrated by the development and application of quantum theory.

Secondly, I want to warn you that we shall be undertaking some experimental investigation together, and one of the experiments will involve a cat of your choice. The investigation will be in the form of thought experiments – that is, devices of the imagination that are used to investigate nature. We will visualise a situation, carry out an operation and see what happens. Undertaking the real experiment may be impossible for practical reasons, but nonetheless we seem able to get a better understanding of nature just by thinking. In fact, the early development of quantum mechanics was largely based on such experiments.

You may question whether mere imagination can have a significant effect on the development of science. Surely it is not possible to learn new things about nature without new empirical data? History resoundingly refutes this objection. Indeed, the most prominent philosopher of science of the twentieth century, Thomas Kuhn, showed that well-conceived thought experiments had often changed widely accepted paradigms.

The experiments teach us something new by helping us to reconceptualise the world in a better way. As we shall see, thought experiments can and have profoundly affected science, technology and the development of the modern world.

Thirdly, at the end of the nineteenth century the most respected physicists of the day thought the process of discovering how the world works was all over. In 1894 Albert Michelson, who was later to win the 1907 Nobel Prize for his experimental prowess, told an audience in Chicago:

It seems probable that most of the grand underlying principles have been firmly established and that further advances are to be sought chiefly in the rigorous application of these principles to all phenomena which come under our notice … The future truths of physics are to be looked for in the sixth place of decimals.

Michelson's views were widely held. Even Max Planck, now considered the father of quantum mechanics, was as a university student advised by Philipp von Jolly, Professor of Physics at the University of Munich, not to study physics with a view to doing research as everything important had already been researched.

As the twentieth century dawned, physicists were firmly of the view that the world was adequately described by theories based on four principles:

1. Physical objects and systems exist separately in space and time and are localisable – that is, they can be generally found in specified parts of the universe.

2. Every event has a cause, and every later state of any system is uniquely determined by any earlier state.
3. All systems evolve smoothly between states.
4. Energy is conserved.

These principles were driving things along rather well. Among other things, they had been responsible for the discovery of electromagnetic waves by Maxwell (1873) and Hertz (1888), and in 1898 J. J. Thomson and Philippe Lenard had used them to discover the small, negatively charged particles we now know as electrons.

Then in 1900 Max Planck, followed by Albert Einstein in 1905, came up with the radical idea that energy exists in small indivisible packets, and thereby triggered a revolution that was to shake both the scientific and philosophical foundations of the world.

The arrival of the theory didn't trigger instant upheaval. Nor, strangely, did it arise from consideration of the new particles and rays being discovered by scientists such as J. J. Thomson, Marie and Pierre Curie, and Antoine-Henri Bequerel. Rather, the key idea of indivisible energy elements arose from the discovery of quite small errors in the theory of the radiation of heat. Imagine, if you will, a metal bar heated to white heat. Clearly, the heat radiated depends strongly on the temperature. What is not so obvious, though, is that the bar glows in not just one colour, or frequency, of light but radiates electromagnetic waves over a wide range of frequencies, of which the light you can *see* is only a small part. The way in which the intensity of the radiation varies with frequency is known as the emission spectrum of the body.

Tom Barnes

As far back as 1859, Gustav Kirchoff had shown theoretically that the emission spectrum depended on the temperature of the body, but he had been unable to predict its mathematical form. Understanding the spectrum was a matter of more than academic interest: the development of the lighting and heating industries of the time depended on it. Hence, theoretical developments were backed up by careful experiments, mainly at Berlin's Physikalisch-Technische Reichsanstalt.

In 1896 Wilhelm Wein made some progress towards solving Kirchoff's emission spectrum problem. On somewhat shaky grounds he proposed a theory for the emission spectrum which appeared to fit well with the experiments that were done through the early and mid 1890s at the Physikalisch-Technische Reichsanstalt. This was where Max Planck came in. The theory of the day considered a radiating body to be a collection of oscillators, each carrying a certain, infinitely variable, amount of energy. In 1899 Planck found an expression for the energy of an oscillator that led naturally to a sound theoretical basis for Wein's radiation law.

Things might have stopped there had it not been for the pesky experimentalists, who in the same year, using improved techniques, found that Wein's law was not in fact correct and gave the wrong results for the radiation of heat at low frequencies. Planck had clearly made a mistake in his derivation of the energy of an oscillator. On realising this, he did something that might be considered somewhat dubious: he guessed a new mathematical expression for the energy that gave the emission spectrum perfectly. He announced this at a meeting of the Berlin Academy of Sciences on 19 October 1900.

However, Planck remained dissatisfied with the complete

Schrödinger's Cat

lack of justification for his guess and reworked the theory. To derive the mathematical expression he had posited for the energy of an oscillator, he was forced to introduce the idea that the total energy of the radiating body was made up of small indivisible elements. He announced this discovery on 14 December at another meeting of the Berlin Academy. Initially Planck didn't think of the introduction of energy elements as a universal quantisation of energy. Nor did he realise that his new radiation law required a break with classical physics. Rather, he saw the idea as a kind of imaginary artefact that enabled him to come up with the correct formula but did not merit serious attention beyond that. What mattered more was the impressive accuracy of the new radiation law.

It was Einstein who, five years later, recognising the revolutionary implications of the quantum hypothesis, proposed that quantisation of energy was a physical reality and developed the theory to describe the properties of energy and radiation. In his 1905 paper, he used thermodynamic arguments to show that, at low energy densities, radiation energy behaved as if it consisted of mutually independent quanta, whose energy was proportional to their frequency. Einstein initially believed that his idea of 'light quanta' was incompatible with Planck's theory but he changed his mind about this in 1906.

As most scientists were convinced that light was a wave, the idea that it consisted of particle-like quanta was radical, to say the least. It did, however, explain some puzzling results that had occurred during the discovery of the electron in 1898. Both J.J. Thomson and Philippe Lenard, then Professor of Physics at the University of Kiel, knew that irradiation of a metal surface with ultraviolet light caused the emission of electrons, and Lenard had made some very rough measurements of their energies. Conventional wisdom had it that the electrons' energy should depend on the intensity of the light on the surface, but Lenard's measurements appeared to indicate that their energy was more related to the light frequency: higher light frequencies ejected more energetic electrons. Lenard never fully understood this, and to make matters worse his experiments were bedevilled by oxidisation of the metal surface in the poor vacuum systems of the time.

Using his new theory, and treating the photoelectric effect as an exchange of energy between an incoming light quantum and an electron at the surface, Einstein was easily able to predict that the energy of the ejected electrons should vary in direct proportion to the frequency of the incoming light. Because of difficulties of the type experienced by Lenard, it took many years for this to be proved experimentally. However, by 1916 a series of experiments performed by the American physicist Robert Millikan at the University of Chicago had shown variation of electron energy in direct proportion to frequency, in line with Einstein's predictions. But none of the experimentalists supported Einstein. Millikan himself described the idea of a quantum of light as 'a bold and reckless hypothesis', and attempted to use classical physics to explain his results. When the classical arguments were shown to be untenable, the photoelectric effect was simply declared to be 'unexplained'. Einstein was simply too radical for his time.

IF LIGHT QUANTA carry momentum as well as energy, it should be possible to use them as a kind of miniature billiard ball. Let's do a simple thought experiment. Imagine a light

Schrödinger's Cat

US physicist Arthur Compton, whose 1922 experiments proved that Einstein had been correct in postulating that light had both wave-like and particle-like properties. AMERICAN INSTITUTE OF PHYSICS/SPL

quantum (or 'photon') interacting with a slowly moving free electron. You might think of this as a kind of elastic collision between two particles. Somewhat the same as an interaction between two billiard balls, the light quantum will bounce back from the electron, which will move off in another direction at increased speed. Total energy will be conserved, but some energy will be transferred from the photon to the electron. Since the energy of the photon is proportional to its frequency, we would then expect the frequency of the photon to drop as a result of the scattering process.

In 1922, Arthur Compton did just this experiment at the University of Chicago. He irradiated a graphite target (which contains free electrons) with X-rays of known frequency. After correction for the effects of relativity on momentum and kinetic energy measurements, he found that the frequency of the scattered X-rays fell by the amount predicted using Einstein's theory: this provided conclusive evidence for the particle nature of photons.

Einstein's prediction that light has both wave-like and particle-like properties presents us with an interesting dilemma. Which picture do we choose, and when? How is it possible for something to have simultaneously the properties of a wave (which is spread out through space) and a particle (whose location is defined within very close bounds)? Let me give you an example to illustrate this. The diffraction of waves is quite easy to observe. Throw a stone into the middle of a still pond and watch carefully how the waves move by any obstructions such as poles and stones. You'll find they don't travel in straight lines. Instead their path bends at each obstruction, and the wave spreads out to more or less fill the space behind.

Light waves behave in a very similar way. Shine the beam from a laser pointer through a pinhole in a piece of tinfoil and you will find that, on going through the pinhole, the light will spread out to form a pattern of rings some distance behind. The intensity distribution in the rings can be calculated exactly using wave theory. On the other hand, if you place a sensitive photodetector in the ring pattern you can observe individual photons, as they liberate electrons in the detector, arriving to create small, randomly occurring, current pulses. The average rate of arrival of photons is proportional to the intensity at the detector.

So does the light act as a wave, or as a stream of particles? The answer is that it acts as both. In simple terms, the intensity distribution calculated from wave theory is a measure of the probability of detecting photons at any point in space, were we to put a detector there. The intensity tells us nothing about *when* a photon will arrive; it is simply a measure of detection probability. Modern experiments which are capable of detecting single photons from the diffraction system show that this theory works well, but it also raises some very important questions, such as:

What happens to the wave when a photon is detected?

Do the photons exist at all until they are detected?

And can light exist simultaneously as a wave *and* a discrete photon?

It wasn't until the late 1920s that debate grew around these questions, and they are still discussed to this day. In the meantime, in the years between 1906 and 1925 considerable progress was made in understanding atomic structure, and despite the atomic models that were developed not being very realistic, this work impinged very much on the development of quantum theory.

You will be aware of Ernest Rutherford's model of the atom, in which tiny, negatively charged electrons are in continuous orbit around a small, positively charged nucleus, and most of the atom is empty space. While at first sight this model appears very logical, it actually has serious defects. The worst comes from the known fact that sending electrons around in a circle – in a synchrotron, for example – causes them to radiate electromagnetic waves. As they orbit the nucleus, the electrons in the Rutherford model should therefore also continuously radiate electromagnetic energy. But if this were the case they would lose orbital energy and eventually (in fact rather rapidly) fall into the nucleus. No atom would be stable.

This problem was partially solved through the development of a kind of semi-classical quantum theory by Niels Bohr, a young Danish physicist who worked briefly in Rutherford's laboratory in Manchester. Bohr's theory of the atom held that only certain electron orbits were allowed, and the energy of an electron in any orbit was quantised – that is, it could take on only certain values set by rather arbitrary rules. Here's how it works. When an electron moves from a high-energy orbit to a lower-energy one, the surplus energy is carried away by an emitted photon, whose frequency is proportional to the energy difference. Similarly, the atom can only absorb energy in quantised amounts when electrons move from orbits of lower energy to those of higher energy. Finally, certain electron transitions are forbidden, leading to lines missing from the spectrum in places where a simpler theory might predict them to be found.

Bohr's theory – for which he received the 1922 Nobel Prize – explained beautifully why excited atoms emit light at particular frequencies only – the so-called atomic line spectra. It also accurately predicted the frequencies of the major lines emitted by the simple one-electron hydrogen atom. However, it broke down very significantly when used to predict the behaviour of atoms more complicated than hydrogen. In the decade after 1913, when the theory was first announced, ever more complicated 'quantum rules' were developed in attempts to explain the details of atomic spectra.

Notwithstanding significant successes, by 1925 the fact that the Bohr quantum theory was unable to explain many important atomic features was generally accepted. In 1925 Werner Heisenberg, who would later receive the Nobel Prize for his work on the uncertainty principle, found a way to reformulate

Schrödinger's Cat

quantum theory so that it was consistent, and not plagued by the difficulties of the Bohr theory. The overriding principle Heisenberg adopted was positivism. The new quantum theory should, he said, be founded exclusively on *observable* relationships between quantities. This criterion was not particularly new or revolutionary but it did involve, for example, doing away with the idea of electron 'orbits'. Rather, the electrons became stationary states which had, associated with them, a certain quantised energy.

Heisenberg described the electron states using arrays of mathematical terms related to phenomena which could, in principle, be observed. Strangely, he found that these arrays of terms did not commute; that is, when they were multiplied together the answer was different depending on the order in which the multiplication was done. It was then quickly recognised that the arrays were in fact matrices, and that Heisenberg's theory could be written in terms of matrix calculus. It also turned out that the non-commutating nature of the matrices was mathematical expression of the uncertainty principle that would later be expounded by Heisenberg as the result of a thought experiment.

Notwithstanding the success of the Heisenberg theory, the matrix calculus was impenetrable to most people, and able to solve only the simplest of situations. Even the one-electron atom proved too complex. A more tractable theory was required. This was provided by Erwin Schrödinger, an Austrian professor of physics at the University of Zürich, and marked one of the key milestones in the development of quantum mechanics. Schrödinger had studied the work of a largely unknown French physicist, Louis de Broglie, who in his doctoral thesis published in 1924 had suggested it was not only light that had a dual

83

Erwin Schrödinger, the Austrian physicist whose 1926 wave theory of atoms led to the idea of quantum mechnical tunnelling, whereby moving particles are able to break through seemingly impenetrable barriers – the basis of many modern electronic devices.
FRANCIS SIMON/AMERICAN INSTITUTE OF PHYSICS/SPL

wave-particle nature: so did matter particles. De Broglie used concepts from relativity and quantum mechanics to show that moving electrons could be considered as waves, with frequency proportional to the electron momentum.

This result would turn out to be valid for all kinds of particles – yes, even for us. And de Broglie's matter waves are diffracted, just like light waves, so next time you drive your car through a narrow gateway, beware: there is a finite probability that you will not go straight through, but be inadvertently diffracted off into the flower-bed! Might I, however, give some advice to parents of teenagers who have done this: the probability of diffraction is very small indeed, so you should not accept the diffraction of matter waves as an excuse for misjudgement of speed or lack of concentration while driving.

Schrödinger's Cat

Schrödinger took the ideas of de Broglie and derived a new wave theory of atoms inspired by wave-particle dualism. The result, published in the German physics journal *Annalen der Physik* in 1926, was a remarkably simple wave equation, in which the momentum and energy of the particle under consideration were represented by mathematical operations on the particle's wave function – that is, its matter wave. Quantisation of the particle energy was not introduced as an axiom, as it had been by Bohr, but arose from obvious matching rules which had to be obeyed by the particle wave function. For example, you can very crudely imagine that there should be a whole number of waves in the equivalent of one electron orbit around the atomic nucleus. Later in 1926, Schrödinger derived a time-dependent wave equation that showed how the wave function of any particle (and therefore both its energy and its momentum) might evolve in both time and space under external influences.

The Schrödinger wave equation is easy to visualise, but its interpretation deserves considerable thought. When Schrödinger first derived his equation he wanted to construct models for particles which consisted of concentrated wave packets, where the particle was represented as a kind of 'pulse' of waves that occurred only in a precisely defined portion of space. However, this proved to be mathematically inconsistent, as waves exist throughout the space in which they are found. Schrödinger then adopted an alternative interpretation in which the particle's wave function (strictly speaking the square of the amplitude of the wave function) is thought of as a kind of density distribution, describing a particle which is in a sense 'smeared out' over a broad region. Clearly this wasn't satisfactory either. For example, we know from cloud chamber experiments that nuclear particles travel well-defined paths in space, and are not smeared out like peanut butter on a piece of toast.

The quandary was solved by another of the greats of quantum mechanics, Max Born. Born suggested that, as with light waves and photons, the wave function simply predicted the *probability* of finding the particle at any point. It provided no guarantee that the particle *would* be found at any particular point in space, but simply indicated the likelihood of detection if one were to place a detector there. We might be able to calculate the wave function of any system perfectly, but there would remain a fundamental uncertainty about the way the system would behave.

As an aside, you may say that theory is all very well but where is the experimental proof that particles behave as waves? Paradoxically, it was J. J. Thomson's son George Thomson who provided this. Working in Aberdeen around 1925, he fired a beam of relatively slow electrons at a thin gold foil and detected a ring-like diffraction pattern on the other side, thus showing conclusively that the electrons were behaving as waves. The fact that J. J. Thomson had showed conclusively in 1897 that electrons were particles, while his son showed conclusively in 1925 that they were also waves, is a particularly apt demonstration of de Broglie's wave-particle duality.

Nowadays, of course, electron waves are used routinely in electron microscopes around the world to examine objects too small to be imaged by light waves.

Schrödinger's wave theory predicts some quite unbelievable phenomena. Let's look, for example, at the scene in the first Harry Potter movie where Harry is looking for Platform 9¾ to board the Hogwarts Express. You might recall that to get on to

Schrödinger's Cat

the platform it was necessary to run as fast as possible into a brick wall – whereupon, if you were lucky, you passed straight through into a magical world. Of course, we all know that is pure imagination, impossible, the product of childhood fantasy. We adults don't believe in that kind of thing, do we?

Well perhaps we should, because the equivalent occurs in the quantum world every day. Write down Schrödinger's equation for the equivalent of Harry and the baggage trolley speeding towards the wall and you will find that the wave function actually exists in the space beyond the wall, albeit with very low amplitude. There is a very small but quite definite probability that Harry and his trolley will make it through the wall. This phenomenon is known as quantum mechanical tunnelling. Quantum mechanics predicts a finite probability that moving particles will be able to pass through barriers which classical physics would call impenetrable. I would suggest, however, that you refrain from checking this out with a trolley full of beer at the local supermarket. The probability of success is vanishingly small and it's almost certain that you will end up in the local lockup with a sore head, rather than safely on the other side of the wall beyond the checkouts.

On the other hand, while you might have a low chance of success at the supermarket, this process happens with high probability in certain situations. The burning of the sun is one. The positively charged nuclei of the hydrogen atoms which fuse together in the nuclear reaction that lights up the sun strongly repel each other, but some manage to get through the barrier. If you use classical physics to calculate the rate at which they come close enough to fuse as a result of thermal motion, you will find this to be far lower than that necessary to explain the

An artist's depiction of the tunnel effect. According to classical physics, a particle can pass through a barrier only if it has kinetic energy greater than a certain critical level, but quantum mechanics predicts there is a finite probability it can do so even with lower energy – a process essential to the nuclear reaction which produces heat and light in the sun. TONY CRADDOCK/SPL

rate of energy production by the sun. The rate at which nuclei fuse can be explained only by the fact that quantum mechanical tunnelling allows the nuclei to pass through the seemingly impenetrable repulsion barrier at a far higher rate than classical physics would predict. In addition, many modern technologies, including electronics, opto-electronics and microscopy, also routinely use tunnelling.

WE NOW HAVE NEARLY ALL the basic building blocks of quantum mechanics. Even at this stage, it is clear that the quantum world is strange, but you haven't seen anything yet. The final component is Heisenberg's uncertainty principle. Earlier, I described how this arose naturally from Heisenberg's matrix theory

Schrödinger's Cat

– which, by the way, was shown to be entirely equivalent to Schrödinger's theory. You might also remember that Heisenberg required his theory to be based on positivism. In keeping with this, he explained the uncertainty principle by means of a thought experiment called the gamma-ray microscope.

It goes something like this. We wish to determine the position of an electron. To do so, we examine it in a microscope that uses very short-wavelength gamma rays. Imagine the electron sitting at the focus of the lens of the microscope. Let's assume that the microscope is so good we need measure only one photon scattered through the lens from the electron to measure the electron's position. I should also point out that diffraction limits the resolution of the microscope – that is, the accuracy with which it can measure the position of the electron is determined by the size of the lens: the bigger the lens, the better the resolution.

Now imagine a photon coming in from the side, scattering from the electron and going through the lens. When this happens, the electron will receive a kick and start moving, just as the electrons did in the Compton experiment. If we repeat the experiment many times, the range of kicks will depend on the size of the lens. With a larger lens (which is better at measuring the electron position) the photons going through it will potentially be scattered through a wider range of angles, and this corresponds to a wider range of kicks imparted to the electron.

You can see that as we try to measure the position of the electron more and more accurately by putting in larger and larger lenses, its momentum will vary over a wider and wider range. In other words, there is a relationship between the uncertainty in the position of the electron and the uncertainty in its momentum. The more accurately we determine the position, the more *uncertain* becomes the momentum. This is true for any particle: we can never simultaneously and perfectly measure both the position and the momentum.

You may think this is simply because the position-measuring process disturbs the momentum and vice versa – a bit like a person with very poor eyesight trying to find a billiard ball on a table using a broken arm. As the person tries to locate the ball, he or she keeps knocking it and moving it, so causing an error in determination of its position. If only they had a delicate enough touch they could find it without moving it. But this isn't actually true, for quantum mechanics tells us that it is not the measurement technique that gives rise to the uncertainty, but rather that the uncertainty is a *property of observation*.

We inevitably ask what this actually means. Heisenberg concluded that it showed the principle of causality, one of the key concepts of classical physics, was incorrect. Even if we know the present state of a system well enough to be able to infer its future state, the uncertainty principle tells us we can never know the present state *perfectly*. Hence our knowledge of the future will be limited. Of course we might imagine that everything is in fact causal, but at some immeasurable level. However this is meaningless, because Heisenberg's view was that those things which are unmeasurable cannot, by definition, be real.

The advent of uncertainty theory also saw the development of the second cornerstone of what was to become known as the Copenhagen Interpretation of quantum mechanics – the complementarity principle of Niels Bohr. We have already seen that it is impossible to observe any system without disturbing it. How, then, can we ever know the state of any system? Quantum mechanics appears to imply that the observer is part of the system being observed, that the classical distinction between the observer and the observed is no longer tenable and, most importantly, that it is fundamentally impossible to obtain objective knowledge. The observer plays a crucial role in quantum mechanics, not just because they always disturb the system under measurement, but because all phenomena are interpreted through their consciousness.

Bohr argued that when we describe the quantum world in classical analogues (that is, in terms of waves or particles) the application of any given analogue automatically precludes the simultaneous application of other equally necessary analogues. The wave picture is in conflict with the particle picture – we cannot use them at the same time in the same experiment. According to Bohr, the physicist must choose which picture to use, because the purpose of physics is not to discover the reality behind the phenomenal world but to predict and coordinate experimental results.

The Copenhagen Interpretation answers questions about the measurement process very simply. In positivist terms, we are unable to say anything about the existence or not of the particle before it is detected, and the detection process is visualised as a kind of collapse or disruption of the wave function, whereupon it probabilistically – and somewhat magically – produces a particle at a given location or, in more complex situations, a system of multiple particles in a given state.

The acausality of nature postulated by Heisenberg and Born was deeply troubling to many physicists, including Einstein and Schrödinger. The difference of views led to one of the most famous episodes in the history of twentieth century physics, the great debate between Einstein and Bohr as to the completeness of quantum mechanics. On philosophical grounds, Einstein denied that the micro-world could be described only statistically. Physical reality, he held, existed independently of man the observer. Bohr's view was, therefore, anathema. In a famous letter to Born in 1926, Einstein wrote:

Quantum mechanics is certainly imposing. But an inner voice tells me that it is not yet the real thing. The theory says a lot, but does not really bring us any closer to the secret of the Old One. I, at any rate, am convinced that He does not throw dice.

Einstein came up with a succession of thought experiments in the hope of demonstrating that the uncertainty relations were not necessarily valid, and that Heisenberg's limit was incorrect. However, for every experiment proposed by Einstein, Bohr was able to provide a counter-argument which showed Einstein's thinking to be untenable.

Einstein didn't give up. Instead, he shifted from trying to demonstrate that the concept of acausality was flawed to demonstrating that the quantum mechanical description of the world was incomplete. To do this, he employed the concept of hidden variables. His argument went something like this:

Quantum mechanics appears to demonstrate that all physical processes have a fundamentally uncertain nature. However, the statistical interpretation of the wave function doesn't necessarily exclude the possibility that individual atomic events might be determined by parameters not yet discovered. The parameters we commonly use – such as position, momentum, energy and time – may be only a subset of those needed to fully describe the physical world.

Schrödinger's Cat

Einstein argued that if quantum mechanics could be reformulated using some kind of 'hidden variables', and if it then reproduced the results of the standard theory, then the known quantum mechanics was probably an incomplete description. There was then no reason to accept the acausality.

Schrödinger shared Einstein's scepticism, and it was his attempt to demonstrate that the Copenhagen Interpretation was absurd that led to the famous thought experiment that has sent shivers down the spines of family cats ever since. In the experiment, a cat is placed in a closed box, together with a phial of poison gas. Also placed in the box is some radioactive material. This is coupled to an ingenious detection system that will smash the phial if a particular radioactive decay occurs. In one hour, there is a 50 percent probability that the diabolical decay will occur, smashing the phial and killing the cat.

The box is left closed for one hour, and then opened again for examination of the state of the cat. Schrödinger argued:

If one has left the system to itself for an hour, one would say that the cat still lives if meanwhile the radioactive event hasn't occurred. The wave function of the system would then have in it both the living and the dead cat mixed or smeared out in equal parts. If it follows from the Copenhagen Interpretation that a cat could be half-dead, and ascribed a wave function that indicates both half-deadness and half-aliveness, is not some other interpretation preferred?

In reality, this experiment is rather trivial and based on false premises. Bohr didn't bother to reply, and even Schrödinger later called it ridiculous, seeing it more as a warning against interpreting the wave function as representing a particle which is 'smeared out' in some way, than as predicting the probability of detection.

Einstein pursued a more scientific argument. In a 1935 paper he published with Boris Podolsky and Nathan Rosen, two of his colleagues at the Institute for Advanced Study in Princeton, he described a thought experiment, now known as the EPR paradox, which attempts to show that quantum mechanics is self-contradictory and therefore incomplete. A rather simplified description of the paradox is as follows:

Imagine a stationary atom which decays, emitting two electrons. After the decay, the atom remains stationary. The electrons must each, then, carry the same momentum and move off in opposite directions at the same speed. This is an example of a so-called entangled state: the quantum states of the two electrons are dependent on each other because they have interacted at some time during the experiment, in this case when the atom decayed.

Now allow the electrons to travel until they are several light-years apart. After this, measure the position of one of the electrons: let's call it electron 1. Imagine you are able to measure the position of the electron exactly.

In making the measurement, you note that Heisenberg's uncertainty principle should apply and so, having measured the position exactly, you can now say nothing about the momentum of electron 1. You can, however, exactly infer the position of the other electron – electron 2 – without measuring it: you know it must be an equal distance on the other side of the decayed atom.

What if you now choose to measure the momentum of electron 2 exactly? Having done this, you will instantaneously know what the exact momentum of electron 1 must be: equal and opposite to that of electron 2. Having made these two measurements, it would appear you then know the position and momentum of both electrons perfectly. In this simple view, the Heisenberg uncertainty principle appears to be violated, and it was on this basis that Einstein, Podolsky and Rosen argued that quantum mechanics must be incomplete.

However, we could get around this by saying that as soon as the position of electron 1 had been measured (and the position of electron 2 inferred), exact measurement of the momentum of either electron would become impossible. Heisenberg would be obeyed. But then there is another difficulty: in making the measurement of electron 1, we have in some way instantaneously influenced what is possible in the measurement of electron 2. The inevitable conclusion is that quantum mechanics predicts some kind of spooky, instantaneous 'action at a distance'. The locality espoused by classical physics cannot be correct.

The concept of a kind of instantaneous action at a distance was originally thought by some physicists to contradict the special theory of relativity, which implies that information cannot be transmitted faster than the speed of light. It is in fact quite easy to show that this is not the case. Even so, we still need some kind of action at a distance, and this requirement can be dispensed with only by assuming the existence of hidden variables.

It wasn't until 1964 that John Bell, a young British scientist working at the European Organisation for Nuclear Research (CERN) in Switzerland, came up with a theory that would lead to a method for determining whether hidden variables were a possibility. Assuming that some kind of hidden variables were present, but not taken into account, in quantum mechanics, Bell calculated the degree of correlation between the measurements made at opposite ends of the EPR experiment. He found that the conventional quantum mechanical theory predicted stronger correlation between measurements – due to the 'action at a distance' effect – than a quantum mechanical theory that might incorporate hidden variables and not require 'action at a distance'.

In 1981, the French physicist Alain Aspect performed an important experiment. He used the two-photon decay of excited calcium atoms to produce pairs of photons whose relative polarisations were known. He then passed the two photons from each pair through separate polarisers, set at particular angles relative to each other, and correlated the signals obtained from them. He found the correlations were as predicted by conventional quantum mechanics, not by hidden variable theory. Variations on this experiment, done since, appear to have confirmed this result.

The inherently probabilistic nature of the universe in which we live has far-reaching implications. For example, if the principles of determinism and causality hold, how does this affect our concept of free will? If we can exactly determine the state of any system, and then predict how that system will evolve using universal physical laws, the concept of free will of the conscious mind cannot, in principle, exist. How, then, can we contemplate holding anyone responsible for their actions? A simple and somewhat cynical view would be that the evolution of the universe is predetermined by nature, and our individual parts in it are nothing more than dancing to a tune set at the beginning of time. Quantum mechanics may provide us with a way out of that dilemma, and allow that choice and responsibility are real.

When Max Planck first announced his theory of quantisation of action over 100 years ago, he can have had no notion of the developments his work would spawn. Yet nearly the whole of modern science and technology rest firmly on his original idea. Even now, we stand on the threshold of a world in which application of the ideas of quantum mechanics will see further massive leaps forward in our technologies, our understanding of the universe, and indeed our understanding of ourselves.

Tom Barnes *is Deputy Vice-Chancellor (Research) and Professor of Physics at the University of Auckland. After completing a PhD in the United Kingdom, he worked in the field of fluid dynamics, and in 1981 came to New Zealand to join the Department of Scientific and Industrial Research (DSIR) in the Physics and Engineering Laboratory, in charge of the optical manufacturing workshop. From 1987 he worked in Japan on optical research, returning to take up a lecturing position at the University of Auckland. He holds several patents, and collaborates in research with Industrial Research Limited and universities in New Zealand, the UK and Japan. He currently serves on the boards of the Foundation for Research, Science and Technology, Auckland UniServices Ltd and the Liggins Institute.*

Journey to the Heart of Matter
—*Paul Callaghan*

IN FEBRUARY 1909, a 35-year-old Irishman, Ernest Shackleton, and his three companions were locked in a desperate struggle for survival as they staggered northwards across the polar plateau, having come within 80 miles of the South Pole before deciding they had to turn back. They had nearly reached the heart of the Antarctic, and the decision to return had been a bitter one as Shackleton knew his failure would allow his arch-rival, Robert Falcon Scott, a chance to succeed where he had failed.

While frozen winds tore at the four men, a much quieter drama was taking place a world away, in a small windowless room in Manchester. There another Ernest, this one 38 years old, was sitting quietly in the darkness with his 20-year-old assistant, peering at a screen that every now and then fluoresced a pinprick flash of eerie green light. It had taken several minutes for their eyes to become accustomed to the blackness. Now they waited patiently, counting every flash and noting down where on the screen it occurred. Each flash of light meant that an alpha particle had hit the screen, an alpha particle that had emerged from a journey to the heart of matter itself, for the two men were in the process of carrying out one of the most famous scientific experiments of all time, a quest to find the structure of the atom and, in particular, what lay at its core.

The older man, who had the previous year won a Nobel Prize, was Ernest Rutherford. The New Zealand-born scientist had discovered the alpha particle while he was a graduate student at Cambridge University. He was to spend the rest of his life learning to understand it, where it came from and how it could be used. His young assistant was Ernest Marsden, a graduate student who could not know then that he would become famous because of the events that were unfolding. Rutherford, though, was very aware of the significance of some of the flashes of light that were occurring far behind the gold foil from which the alpha particle had scattered. It was what he had secretly hoped for, the very essence of the design behind his experiment. Backscattered alphas – alpha particles that had been turned around in their tracks – could mean only one thing: at the heart of the atom was a dense, heavy nucleus.

Rutherford picked up a pencil and started to calculate, balancing the kinetic and electrical energies behind the processes. He knew this would tell him how big the nucleus was. His hand trembled a little as he reached the answer, the pencil trailing illegibly on the page in his notebook as his mind raced. He stared at the number he had written, and for a moment said nothing. What he had discovered was breathtaking: the nucleus was some 10,000 times smaller than the atom itself. Rutherford had seen into the atom as plainly as if he had been on a personal journey to the interior and viewed the nucleus with his own eyes.

Rutherford turned to Ernest Marsden, passing him his notebook and pointing to the end-point of his arithmetic. Marsden looked carefully, and then checked Rutherford's calculations. They were right. His heart pounded. In sharing that moment, an irrevocable bond had been forged. Ernest Marsden had connected with his hero in a way that would change his life, and ultimately cause him to develop deep links with Rutherford's homeland.

Journey to the Heart of Matter

By March 1909 Shackleton and his companions had made it back, first to the shelter of their hut on Cape Royds, then to Lyttelton, New Zealand, where Shackleton sent telegrams before the men embarked on the voyage back to Britain. The heart of the Antarctic would not be reached for three more years, and over those three years science and technology would change forever because of Rutherford's and Marsden's new knowledge of the nature of matter. A quest that had begun in Greece over 2000 years before had ended, and the modern age had truly begun.

With the discovery of the atom and its structure, the whole of chemistry was suddenly transformed. At last chemists knew how chemical reactions took place, making it possible to develop new materials – light, strong, high-performance polymers, colourful and comfortable fabrics, life-saving pharmaceuticals. It was also now possible to control the properties of silicon and other semiconductors by using atomic dopants, giving us solid-state electronics, lasers, computers and the digital era. The atomic age also gave us the measurement tools of atomic spectroscopy, magnetic resonance, and electron and X-ray diffraction, letting us probe the structure of proteins and DNA, to start to understand how life works, the nature of disease and how to treat it.

Life goes on, as we say, but its forms alter. The cells in our body are recycled, most lasting no longer than a few months, with the exception of our precious brain cells, which we hold for life. But eventually the span of all living creatures comes to an end and we die, even if our genes survive many more generations before they, too, mutate. Only our atoms remain, unperturbed, shared by each successive generation and every species on the planet.

Ancient Greek philosopher Democritus was the first to propose that all matter was made up of indivisible elements, which he called *atomos*. This seventeenth century drawing depicts the Democritean universe, with Earth and the planets at the centre, surrounded by the starry heavens, and beyond that an 'infinite chaos' of atoms. CORDELIA MOLLOY/SPL

The story of atoms began in Greece around 400 BC, when a philosopher called Democritus gave the world the idea that all matter was made up of eternal, unchangeable elements, for which he coined the word *atomos*, meaning indivisible or uncuttable. After Democritus, the story of the atom stalled until 1738, when a Swiss physicist, Daniel Bernoulli, showed he could explain the pressure exerted by a gas on the walls of its container by thinking of the gas as minute particles with an energy of motion associated with the amount of heat in the gas. Bernoulli called the particles atoms, but we now know, of course, that if a gas comprises molecules made up of more than one atom, it is the molecule that is doing the colliding. This confusion between atoms and molecules was to dog science for another century.

French chemist Lavoisier is depicted showing fellow scientists his experiment to find the composition of air. He discovered the gas oxygen and proved it was hard to break down by normal chemical processes, but his job tax-collecting for the royals led to his execution in 1794. SPL

Today we know that the atom is the smallest unit of the chemical element, a substance that cannot – as Robert Boyle, Antoine-Laurent Lavoisier and Joseph Priestley discovered in the eighteenth century in their efforts to understand the element oxygen – be decomposed into simpler substances by ordinary chemical processes. Boyle and Priestley had the good fortune to be English and well-educated, but Monsieur Lavoisier was French and well-educated – and had royalist connections. For him, in the late 1700s, that meant the guillotine, but not before he had helped give us our understanding of elements.

MANCHESTER IS ONE OF ENGLAND'S more impressive cities. Its elegant commercial buildings, library and town hall all speak of a former age of grandeur. It was in Manchester in 1909 that Rutherford carried out his famous experiment, and in

Manchester a century earlier that the role of atoms in chemistry had first been suggested. In 1805 John Dalton, secretary of the Manchester Literary and Philosophical Society and a public and private teacher of mathematics and chemistry, was studying the various gases that could be made by combining oxygen and nitrogen. He found that the two elements united chemically in simple numerical ratios to form compounds: nitrous oxide in the ratio 2 to 1; nitric oxide in the ratio 1 to 1; and nitrogen dioxide in the ratio 1 to 2. In 1808 Dalton published a paper in which he stated that the rule for calculating the proportion could be based on the idea of atoms, with all atoms of an element being of exactly the same size and weight, and in these two respects unlike the atoms of any other element. If you visit Manchester you will see Dalton's bust at the entrance to Manchester Town Hall. Sadly his records, carefully preserved for a century, were destroyed during the bombing of the city in World War II.

The first person to connect Bernoulli's idea of molecules in a gas with Dalton's idea of atomic weights was an Italian, Amadeo Avogadro. An ecclesiastical lawyer who had developed a strong interest in mathematics and science, Avogadro in 1811 made the suggestion that eventually allowed for the counting of atoms. He hypothesised that equal volumes of different gases would contain exactly the same numbers of atoms when under the same conditions of pressure and temperature. The particular mass proportions which Dalton found in his chemical reactions gave a unit of mass, known as the mole, for each element, an amount which always contained the same number of atoms. We now call this number Avogadro's number.

To many chemists, however, these 'atoms' were really fanciful concepts. They provided a set of rules for predicting chemical

Journey to the Heart of Matter

reactions, but that didn't mean they were real. And to some physicists the idea that gas pressure resulted from the collisions of atoms at the wall of the container was not even a useful rule: it was not only fanciful but in a sense immoral, given that atoms could not be 'seen'.

The fiercest and saddest battle over the existence of atoms and molecules would rage in Germany and Austria. The person at the centre was the Austrian physicist Ludwig Boltzmann, a rather awkward man, ambitious but given to self-doubt and introspection. Boltzmann argued that temperature was nothing more than a reflection of the thermal energy of atoms and molecules in a gas, the restless random motions which caused them to exert pressure on a surrounding container. He suggested how much thermal energy should be present in every type of motion, and showed he could use this to explain the laws of thermodynamics – laws that were exactly obeyed, and whose formulation was based on the ways in which heat and work manifested themselves in gases. These were the basic laws that would come to govern steam engines, and for that matter jet engines and internal combustion engines.

Boltzmann had a hard time. There were several objections to his ideas. The first was that if one used the mechanics of gas molecules to explain gas laws, then, since the laws of mechanics were reversible, the laws of thermodynamics should also be reversible. That would mean that heat could flow from a colder temperature to a hotter temperature under its own volition, something never observed. Boltzmann easily rebutted this objection, explaining that while such reversion was possible, it was extremely improbable given the huge number of atoms taking part. However, he said, in systems with small numbers

Austrian physicist Ludwig Boltzmann, whose theory on the thermal energy of atoms and molecules would be used by Einstein in his 1905 paper on Brownian motion.
SEGRE COLLECTION/AMERICAN INSTITUTE OF PHYSICS/SPL

of atoms, reversal of the law might be seen – something we now know to be true. But Boltzmann's reasoning was way ahead of its time and most found it too hard to comprehend.

Another Viennese physicist, Ernst Mach, after whom the speed of sound is named, went further, saying it was distasteful to speak of atoms and molecules when no one had seen them and there was no way of directly verifying their existence. To Mach, scientific laws had to be based on what was directly measurable, not on fanciful theory. His view may well resonate with people of common sense, but common sense gets us only so far. Science is, as we have learned, a method of discovering truths that defy common sense. It is, after all, common sense that the sun revolves around the Earth.

The disagreement between Mach and Boltzmann was at the heart of one of the greatest struggles in physics, the build-

Journey to the Heart of Matter

ing of theoretical physics, a branch of physics that burrows deep below the surface, coming up from time to time to test its predictions. It was a struggle Mach was eventually to lose, and the definitive end would come in 1905, when the work of a new theoretical physicist called Albert Einstein vindicated Boltzmann completely. Sadly, it would be too late: Boltzmann died by his own hand the following year, in a state of deep depression and not knowing of the Einstein paper that had proved him right.

The paper was one of the five published by Albert Einstein in 1905. It was on Brownian motion, the curious jiggling about of pollen grains observed under the microscope in 1827 by the botanist Robert Brown. Einstein boldly postulated that a particle such as a pollen grain, which was large enough to be seen in an optical microscope, could still be small enough to be buffeted by the random thermal motion of the surrounding water molecules. Einstein used Boltzmann's predictions of the amount of thermal energy in the motion of atoms and molecules to come up with a rule by which, if one accepted his theory, one would be able to estimate Avogadro's number simply by observing how far the pollen grain moved over a given time. His experimental work was completed in 1908 by the French physicist Jean Perrin, who won the Nobel Prize for it.

With Einstein's explanation of Brownian motion and Perrin's accurate measurements, it became possible to calculate not only the size of the atom, but Avogadro's number. It turned out to be vast: greater than 6×10^{23}. Atoms were, therefore, very small and very numerous. Mind-bogglingly small, and mind-bogglingly numerous. If we could zoom you so your body was the size of New Zealand, with your head at Cape Reinga and your feet at

Bluff, one of the atoms in your body would be the thickness of a human hair – in other words you could just see it.

How numerous? In your body, the number of molecules of water is 10^{27} (1 with 27 zeroes after it). Each molecule has two atoms of hydrogen and one atom of oxygen, and so the number of oxygen atoms is also about 10^{27}. Molecules are clusters of atoms bound together. They can range in size from as small as a couple of atoms to huge molecules such as DNA, containing millions of atoms.

Think about that 10^{27} oxygen atoms in your body. Given that the number of people in the entire world is only 10^{10}, we can draw some remarkable conclusions about how many atoms we get to use when they are shared about. Take any historical figure who lived sufficiently far in the past that his or her atoms have long since entered the biosphere and been shared about, as ours will all be eventually. It doesn't matter who your historical hero or heroine is, if they died a few hundred years ago you will almost certainly have a few million of their atoms in your body, a fact which alone should convince you that atoms are indeed very small and numerous.

Although atoms and their combinations – molecules – came to be accepted, no one knew what an atom was made of. The answer would be uncovered in 1909 by the New Zealander Ernest Rutherford. Rutherford was born in Brightwater near Nelson in 1871. When he was eleven he moved with his family to the town of Havelock, where he came under the influence of a remarkable primary school teacher called Jacob Reynolds. In 1887, after two attempts, he won a scholarship to Nelson College, where he excelled in physics, maths and chemistry. In 1889, again after two attempts, he won a scholarship to Canterbury College, where he met the second great academic influence in his life, the eccentric and colourful Professor Alexander Bickerton. Rutherford soon discovered that his strength was in experimental physics, and as part of his Masters thesis he developed an ingenious, highly sensitive magnetic detector of high-frequency oscillatory fields, or what we now call radio waves.

Rutherford seems to have been destined to be a scientific researcher. In 1895, after a candidate higher on the list withdrew, he won the 1851 Exhibition scholarship, which took him to Cambridge University as one of its first-ever PhD students. There he was to meet the third great influence in his life, J.J. Thomson, the head of the Cavendish Laboratory and discoverer of the electron. Rutherford astounded his Cambridge professors with his experimental skills, his radio-wave detector being far more sensitive than anything they had seen.

It is not known why Rutherford decided to leave radio research and take up the study of atomic phenomena. Perhaps it was the influence of Thomson, who was keen to use the new X-rays discovered by the German physicist Wilhelm Conrad Röntgen the year Rutherford arrived in Cambridge. Whatever the reason, we can be glad Rutherford shifted fields. While a student in Cambridge, he was to discover alpha rays, show they were particles, and measure their charge-to-mass ratio. These alpha particles, which emerged from radioactive materials extracted from radium- and uranium-bearing rocks using the methods of Marie Curie, were to be the basis of his remarkable future achievements.

According to Rutherford's New Zealand biographer Dr John Campbell, Ernest Rutherford was never quite comfortable at Cambridge, and on finishing his PhD he left for a job at McGill

University in Montreal, Canada. His time at McGill from 1898 to 1907 was to be wondrous, bringing him in 1908 a Nobel Prize for his discovery of the laws of radioactivity. That year he accepted a position as Professor of Physics at the University of Manchester. Here he returned to studying alpha particles emitted from the element radium. By capturing the particles in a glass vessel and neutralising their positive charge, he and his student Thomas Royds were able to show that they converted to the element helium. Clearly, whatever these charged alpha particles were, they were closely related to this second lightest element.

There is no doubt, though, that Rutherford's brainwave of firing the tiny alpha particles at a gold foil so as to probe the nature of atoms themselves was his most audacious. It was known that atoms were electrically neutral but contained within them the tiny, negatively charged electron which Thomson had discovered in 1897. But if atoms were electrically neutral, there had to be positive charge within the atom to balance the negative charge of the electrons. And something had to give the atom its mass, the electron's mass being some thousands of times too small to explain atomic mass. Whether Rutherford had a hunch that the mass and positive charge were concentrated in a particular part of the atom we cannot be certain, but his use of alpha-particle 'scattering' from gold was creative experimental physics at its best. Later, he was to say of the alpha particles scattering backwards: 'It was almost as incredible as if you fired a 15-inch shell at a piece of tissue paper and it came back and hit you.' The statement had an element of theatricality: Rutherford's whole experimental design suggested he had expected the result.

Journey to the Heart of Matter

What had scattered the alpha particle? It had to be an object more massive and more highly charged than the particle itself. Rutherford's inspired guess was that this large mass and charge were concentrated in a very small object, the nucleus of the atom. It is sometimes said that Rutherford was just an intuitive experimentalist and no good at the mathematical aspects of physics, yet he derived a formula which showed how alpha particles would scatter if such a nucleus existed, and his calculations exactly matched the scattering pattern he and Marsden had observed. Rutherford had discovered the nature of the atomic interior. Most of the mass and all of the positive charge were concentrated in a region of space tens of thousands of times smaller than the atom itself, while the delicate electrons carried the negative charge in surrounding orbits. It now became clear that the alpha particle itself was the nucleus of the helium atom – the mass and the two units of positive charge without the electrons.

It was a young Danish physicist, Niels Bohr, who, when visiting Rutherford's Manchester laboratory in 1911, saw that this new model of the atom could be used to explain the mystery of atomic spectra, the distinct wavelength of light which came out of atoms when they were heated up. Einstein had suggested that light interacted with matter in the form of little packets of energy, whose size was determined by their wavelength. Bohr speculated that if the electrons were confined to particular orbits, changes in those orbits would give the discrete energies needed to explain the special wavelengths. The existence of special orbits gave rise to quantum theory and the new science of quantum mechanics, by which the details of the atom could finally be described with complete accuracy. In a few years in the early twentieth century, the whole issue of atoms had been settled.

Niels Bohr (left) with Max Planck. Planck's 1900 quantum theory, together with Einstein's theory of relativity, revolutionised our understanding of matter, space and time. Bohr's idea that electrons had special orbits allowed atomic structure to be, at last, fully explained.
MARGRETHE BOHR COLLECTION/AMERICAN INSTITUTE OF PHYSICS/SPL

It was left to Rutherford's brilliant student and protégé Harry Moseley to show that the positive charge held in the nucleus of the atom gave an element its position on the chemical periodic table – what we call the atomic number. With this addition to Rutherford's discovery, the foundations of modern chemistry were truly laid.

In 1917 Rutherford carried out his final experiment of genius. Instead of firing the alpha particles at the heavy gold atom, he fired them at the much lighter nitrogen atom. The nitrogen nucleus absorbed the alpha particle and then, by emitting a hydrogen nucleus (also known as a proton), turned into a nucleus of oxygen. Rutherford had not only discovered the atom, he had split it asunder, showing that even atoms were

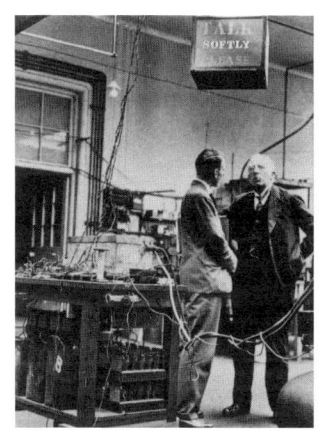

Ernest Rutherford (right) in the Cavendish Laboratory at Cambridge University. Two years after splitting the atom at Manchester University in 1917, Rutherford was appointed director of the Cavendish and returned to the laboratory where he had researched atomic structure as a PhD student 22 years earlier. PROF. PETER FOWLER/SPL

not entirely immutable. His success recognised with a knighthood, he at last returned to Cambridge, this time as Cavendish Professor of Physics and head of the Cavendish Laboratory. It was the beginning of the years of triumph of Cambridge physics, when Rutherford showed that he was as much a genius of science organisation and leadership as of the art of experimental physics. That leadership centred on his ability to inspire others and then quietly and generously support them. There are many stories about his eccentricities, such as not letting his students work too long in the laboratory. One evening he inquired of a student he had noticed working particularly hard: 'Do you work in the mornings too?' 'Yes,' the student answered proudly, sure that he would be commended. 'But when do you think?' Rutherford said.

PAUL CALLAGHAN

Rutherford was certainly kinder to his students than Einstein was. There is a famous, and possibly apocryphal, story of one of Einstein's students coming to him and saying: 'The questions in this year's exam are the same as last year's!' 'True,' Einstein is said to have replied, 'but this year all the answers are different.' Not surprisingly, Einstein had few research students. By contrast, Rutherford sup-ervised a dynasty of future scientific greats: nine of his students went on to win Nobel Prizes.

And what of young Ernest Marsden? Encouraged by Rutherford, he moved to New Zealand in 1914 to become Professor of Physics at Victoria University College. Soon afterwards he left to serve as a signals officer with the New Zealand Expeditionary Force in France, where he was awarded the Military Cross. Marsden was lucky: he survived. Harry Moseley was killed by a sniper's bullet at Gallipoli, a grievous loss to British science of one of its brightest and best. Rutherford was devastated by his young colleague's death.

After World War I, Marsden joined Rutherford for a period in Manchester and then returned to New Zealand as Assistant Director of Education. In 1926 he became the first head of the Department of Scientific and Industrial Research. Rutherford, meanwhile, went on to become president of the Royal Society of London and a peer. He died in 1937 from a strangulated hernia – an unnecessary death probably caused by the delay getting up from London a Harley Street surgeon of suitable status to treat a lord of the realm. John Campbell described it as 'dying of fame'. Campbell's biography is one of 42 written about Rutherford, making him the most famous New Zealander who ever lived.

Not everyone welcomed the new discoveries about the nature of matter and life. In 1933 Einstein was forced to flee

Ernest Marsden, whose experiments with Ernest Rutherford while a graduate student at Manchester University would make his name, later emigrated to New Zealand and in 1926 became first director of the DSIR (Department of Scientific and Industrial Research). SIR C. FLEMING COLLECTION, ALEXANDER TURNBULL LIBRARY, F-18564-1/4

Hitler's Germany, ultimately seeking refuge in the United States. One hundred Nazi professors subsequently published a book condemning his theory of relativity. 'If I were wrong,' Einstein wryly remarked, 'one professor would have been enough.'

Manchester University threw out Rutherford's apparatus and converted his laboratory into a teaching space for psychology. In New Zealand, we tore down the house in which he was born and no trace of it remains. Belatedly, the country has made amends. There is a memorial to Rutherford near his birthplace at Brightwater near Nelson, and the Canterbury Arts Centre houses Rutherford's den, the actual room in Canterbury College where he conducted research for his Master of Science degree in the 1890s. You can find Rutherford on New Zealand's

100-dollar banknote, along with the graph of radioactive decay which won him his Nobel Prize. Rutherford used the law of decay to find the age of the Earth, the answer coming out at 4500 million years.

Now, at the beginning of the twenty-first century, using the new tools of electron microscopy, scanning tunnelling microscopy and atomic force microscopy, we can see atoms directly and manipulate them one by one. The era of nanotechnology has arrived, making possible more powerful and smarter electronics, and new devices for the treatment of disease. We can interact light photons with atoms, making it possible to form the building blocks of the quantum computer that will revolutionise the information age. We can use the atomic tools of magnetic resonance and radiation scattering to understand the details of bio-molecules, how life works and how cells function.

But one of the most remarkable legacies of Rutherford has been the knowledge that the atoms that make up the Earth are not primordial. The primordial matter, hydrogen, fuses to form helium in the nuclear reactions of stars. Carbon, phosphorous, nitrogen and oxygen, the atoms of life, are formed in these nuclear reactions. Iron, lead, uranium and the other heavier elements that make up the Earth can be formed only in a supernova explosion, the spectacular death of a star. All of us here on Earth are, then, recycled stardust, atoms left over from the life and extraordinary death of a star long gone from our universe. That's as romantic an idea as Shackleton could ever have wished for.

Paul Callaghan was born in Wanganui, New Zealand. He obtained his DPhil degree from Oxford University, working in low-temperature nuclear physics. On his return to New Zealand he began researching the applications of magnetic resonance to the study of soft matter at Massey University, and in 2001 was appointed Alan MacDiarmid Professor of Physical Sciences at Victoria University of Wellington. He also heads the multi-university MacDiarmid Institute for Advanced Materials and Nanotechnology. Paul Callaghan has published around 220 articles in scientific journals as well as a book on magnetic resonance. In 2001 he became the thirty-sixth New Zealander to be made a Fellow of the Royal Society of London. He is a founding director of magritek, a small company based in Wellington, which sells NMR instruments.

The Unconquered Sun
—*Robert Hannah*

IT IS A TRUISM TO SAY that time is fundamental to physics. How, though, was time measured in the ancient world – and in particular in those parts of the ancient world to which we owe our concepts and practices of timekeeping? To find out, we need to transport ourselves imaginatively as far back as 4000 years, to a region that stretched from the Middle East to the Mediterranean Sea.

This isn't as hard as it may sound. Cast your mind back a few years to when plain drivers' licences were transformed into photographic ones. We were told we had to upgrade our old licences to the new type within 60 days after our birthday. But there was also a less formal version of this message, a monthly television advertising campaign telling us to use astrological star signs as the signal to upgrade. Taurians were to renew when the sun was in the constellation of Taurus, Virgoans when it was in Virgo, and so.

Adopting the signs of the zodiac as a calendar like this was one of a number of mechanisms ancient peoples used for keeping time. It shared with other early methods a common principle: the observation of celestial bodies – the sun, the moon and the stars. We need to understand some of the astronomy that underlies these observations, and from an ancient, Earth-centred perspective. (Our everyday language still retains this perspective. For example, we talk of 'sunrise' and 'sunset' even though we know this is not an accurate description of what is

As the Earth rotates around its own axis, the stars appear to move in circles, as seen here in a long-exposure photograph of star trails around the south celestial pole, taken from the centre of Stonehenge Aotearoa, an outdoor observatory in rural Wairarapa, New Zealand.
IAN COOPER

happening.) From our viewpoint on Earth, we see the celestial bodies move, in general, from east to west. The sun rises in the east and sets in the west, more or less. The moon rises and sets in the same way, and travels along much the same path as the sun.

The stars share this east-to-west motion, although not all rise and set. Why is this?

We know now that as the Earth rotates each day around its own axis, it creates for us the sense that celestial bodies are rising and setting: we see the stars, in particular, move from east to west in parallel semicircles above the horizon. These semicircles are really full circles, continuing under the horizon as the Earth spins. The axis around which the stars seem to wheel is the axis of the Earth extended out into space. Exactly like the lines of latitude on a globe, these circles are smallest at the northern and southern poles of the extended axis, and largest at its midpoint, or equator, which is simply the extension of the Earth's equator out into space too.

For people who live closer to the equator than to the North Pole – as is the case for people in the Mediterranean and Middle East – the horizon cuts across the circular paths of the stars in such a way that stars close to the northern pole in the sky will not rise or set, but will appear to circle perpetually around the pole. For these same observers, stars closer to the south celestial pole will not even rise above the horizon, but will always remain invisible. Just which stars will always stay above the horizon, which will rise and set, and which will never be seen depends on the particular latitude of the observer on Earth, and can be readily calculated if that latitude is known. However, experienced long-distance travellers in antiquity would have gained a 'road sense' of when certain stars would dip permanently out of sight as they travelled north, or which would appear anew from below the horizon as they travelled south.

There are also larger cycles than just the daily ones of rising and setting. From one week to the next, the moon presents a different phase, starting 'new', growing bigger through first quarter to 'full', and then getting smaller from full to last quarter and back to new. And if you are a reasonably keen observer of the night sky, you will notice that the stars visible in one season of the year are different from those visible in another. Fortunately, we don't have to understand why this is so in order to understand ancient systems of telling the time. The basics are simple: the sum of light and dark gives us the day; the phases of the moon provide us with the month; the changing seasons introduce us to the year. It's not surprising, then, that these three categories of celestial phenomena – solar, lunar and stellar – lie at the heart of most calendars.

Let's start with the sun. Over the course of a year, as long as we don't impose artificial changes such as daylight saving, we see day and night gradually change in length. In midsummer the days are at their longest, while nights are shortest. In midwinter the opposite applies. Halfway between, day and night are practically equal in length. The midsummer and midwinter points are called the solstices, and the points halfway between are the equinoxes. For ancient peoples, the midwinter solstice was a focal point in the agricultural year. From then on the nights would get shorter and the days longer and warmer, so planting could take place. Because daylight started to grow again, midwinter could be taken as the annual birthday of the sun, and in fact the Romans celebrated the Birthday of the Unconquered Sun at midwinter, which in their calendar was 25 December.

From one season to the next we see the sun apparently shifting north or south along the eastern and western horizons. In midsummer, we see the sun rising or setting at its most southerly points. As we move into autumn and then winter, its rising or setting point on the horizon shifts as well, moving further and further north, until in midwinter it reaches its most northerly point. After that, the sun returns along the track it has measured out on the horizon to the midsummer point. If we were in the northern hemisphere, where most ancient civilisations existed, the midsummer point would be the most northerly and the midwinter point the most southerly.

The Unconquered Sun

The prehistoric monument Stonehenge on England's Salisbury Plain. Was it an observatory, a timekeeper or a place-finder? SACHA HALL

In the late nineteenth century, the Hopi Indians of Arizona were still watching the sun's shifting position on the horizon to help find the time for their midwinter celebrations. When the chiefs responsible for the ceremonies saw the sun set in a valley between the mountains 100 kilometres away, they knew they needed to begin the nine-day festival in four days' time. Towards the end of this festival the winter solstice would occur.

This movement of the sun, and the closely related movement of the moon, are also thought to have been used for orienting tombs, henges, and cursus monuments – long, narrow earthwork structures – in northern and western Europe in the Neolithic period of 4000–2000 BC. Stonehenge is the best known example of this type of monument – an inner ring of large stone pylons encircled by an outer ring of holes, presumably to carry markers of some sort, all inside a ditch. Outside stands a large stone known as the heel stone. If you stand in the centre of the circle and look towards this stone, you will see the point on the horizon where the midsummer sun will rise.

Based on this, some people have claimed that such henges were observatories, from which astronomical observations of the sun and moon were made and lunar eclipses predicted. But there's no incontrovertible evidence. The fact is that some orientations are inevitable. If Stonehenge is oriented through the heel stone to a point on an otherwise featureless horizon where the midsummer sun rises, it is a useful fact of cosmic geometry that opposite that point the midwinter sun will set. However, this tells us nothing of why the monument may have been oriented in this way. And we don't know what took place at the henge, information which might help explain its apparent orientation. Perhaps, rather than being timepieces, the henges were place-finders. It may be that, as a recent study attempted to show, circular monuments could be used to determine latitude.

In the absence of evidence, over-interpretation is always a problem. You may know the story of Sherlock Holmes and Dr Watson on a camping trip. After a good meal and a bottle of wine, the two men lay down for the night and went to sleep. Some hours later, Holmes woke up and nudged his friend.

'Watson,' he said, 'look up and tell me what you see.'

Watson replied, 'I see millions and millions of stars.'

'What does that tell you?' asked Holmes.

Watson pondered the question for a moment. 'Astronomically,' he answered, 'it tells me that there are millions of galaxies and potentially billions of planets. Astrologically, I observe that Saturn is in Leo. Horologically, I deduce that the time is approximately a quarter past three. Theologically, I can see that God is all-powerful and that we are small and insignificant. Meteorologically, I suspect that we shall have a beautiful day tomorrow. Why, Holmes, what does it tell you?'

The Unconquered Sun

Holmes was silent for a minute, then he spoke. 'Watson, you idiot, some fool has stolen our tent!'

Other Neolithic structures demand attention because we know their purpose, and large numbers of these are organised in specific directions. Many prehistoric tombs in western Europe face in clusters towards where the sun rises or sets, or towards the sectors of the sky where it climbs or declines. Of several thousand of these tombs, only a few face north – to where, in the northern hemisphere, the sun never crosses. Why this is, we don't know. A simple explanation is that most of the tombs face towards where the sun rose or set on the day construction began.

Interestingly, early societies in southern Europe and the Middle East were not as focused on the sun's position on the horizon. The reason is worth emphasising, because it's perhaps not obvious now. For places in the south, the sun travels on a much smaller arc along the horizon from one solstice to the next than it does in more northerly latitudes. This means that its change of position is too small and gradual to be distinctly pinpointed on a daily basis. But as we travel north, the size of the arc increases noticeably. For example, at the site in Iraq of ancient Babylon, 32 degrees north, this arc extends just 56 degrees from the summer solstice down to the winter solstice. But in Uppsala in Sweden, at 60 degrees north, where there are barrow tombs oriented towards the sun, the arc is a whopping 102 degrees – more than half the eastern horizon and almost twice as large as the arc at Babylon. This makes the job of tracking the sun day-by-day much easier in northern latitudes because the sun moves over a more measurable distance. In the month leading up to the spring equinox, the sun shifts just 12 degrees along the horizon in Babylon, but 21 degrees – almost twice the distance – in Uppsala.

This is why people in southern Europe and the Middle East wouldn't have been attracted to using the sun on the horizon to tell the time of year. Instead, they turned to the stars. As early as 1000 BC the Babylonians developed a calendar which included observations of the risings of certain stars at dawn and dusk through the year. One calendar compiled in the seventh century BC from much earlier star catalogues, and called *Mul.Apin* ('The Plough') after its opening line, tells us that:

> *On the 1st day of the month of Nisannu,*
> *the Hired Man [Aries] becomes visible.*
> *On the 20th day of Nisannu, the Crook*
> *[Auriga] becomes visible.*
> *On the 1st day of the month of Ayaru, the Stars*
> *[Pleiades] become visible.*
> *On the 20th day of Ayaru, the Jaw of the Bull*
> *[Hyades] becomes visible.*

The Maori calendar also uses this method. It begins the year at the first dawn rising of the Pleiades, a faint cluster also known as the Seven Sisters, which takes place in the middle of June. This drove the ethnographer Elsdon Best to wonder if there were a connection between Maori and the peoples of ancient Babylonia, since both used the Pleiades for their calendars. Such a connection seems highly unlikely, but use of the Pleiades in the South Pacific is intriguing because these stars are not at all easy to see here. They skim the horizon, rather than climbing high into the sky as they do in the Middle East. This suggests that the Maori calendar was created in a more northerly region of the Pacific than New Zealand.

The Unconquered Sun

The ancient Greeks also used observation of the stars to signal the seasons, with the earliest records, in the poems of Homer and Hesiod, dating back to between 750 and 700 BC. In his famous rural epic 'Works and Days', Hesiod used the stars to advise farmers when to do various jobs, as in this example:

> *When Pleiades and Hyades and the strength of Orion set, then remember it is the season for ploughing …*

Having mastered the seasons, the next challenge was to define time from day to day. This type of star calendar occurred in Athens just before 400 BC in an instrument called a parapegma, a large set of stone tablets inscribed with 365 holes, one for each day of the year. Beside many of the holes were added the star observations for the day. Someone then had the job of moving a peg from one day to the next through the year. A parapegma compiled in Athens around 432 BC and reconstructable from later texts probably looked like this for late July/early August (O indicates peg-holes):

> O *Sirius is visible. The stifling heat begins.*
>   *The weather changes.*
> OOOOOOOOOOOO
> O *The heat is at its greatest.*
> OO
> O *Lyra sets. It also rains. The Etesian winds stop.*
>   *Pegasus rises.*

This calendar, or almanac, also mentions the solstices and equinoxes, and even marks the points in between, which today we call 'Quarter' days. These Quarter days often served as markers for farming activity and the start of the agricultural seasons, and still do so in Scotland, for instance. In the example, the ceasing of the Etesian winds marked the end of summer and start of autumn. While the winds blew, the summer's harvest of wheat could be winnowed. When they stopped, it was time to move on to picking the grapes.

As a timekeeping device, the risings and settings of stars have a very long, continuous history. By the Middle Ages, Christian monks were using the stars to signal the hours for evening prayer. In a stripped-down form, the practice survives to the present day in Western astrology's use of the zodiacal signs, such as Aries, Taurus and Gemini, to indicate to individuals when they should (or should not) take certain actions.

The zodiac consists of the stars through which the sun appears to track during the course of the year. By about 500 BC these stars had been organised by the Babylonians into twelve constellations. The sections of the sky occupied by the constellations were then artificially formed into signs of equal size – 30 degrees – over the next 200 to 300 years. The names of these signs that have come down to us are Latin translations of earlier Greek names. The Greek names, in turn, were translations of the original Babylonian names.

Soon after 400 BC, it seems, the Greeks devised a type of sundial which showed the passage of the sun into each of the zodiacal signs through the course of a year. We can imagine how helpful this type of instrument would have been for marking out the passage of the seasons, and even for dividing the year into smaller, more manageable chunks – the twelve zodiacal months. But how useful would it have been on a day-to-day basis? Unfortunately, at the latitude of Athens in Greece the shadow measured by a pointer changes only a little from day to day, too little to help in the organisation of events from one day to the next. However, the movement of the sun across the sky in the course of the day was more distinct, and therefore offered a better opportunity to organise events *within* the day, and we can see later sundials taking much more advantage of this.

Sundials were an enduring feature of the ancient world. The earliest surviving ones, which come from Egypt and date from the fifteenth century BC, consist of an L-shaped block of wood. The long 'back' of the L was meant to be laid out horizontally, with the much shorter 'foot' standing upright to face the sun. A shadow was cast back on to the horizontal block, which was marked out with a scale of hours.

Where did these 'hours' come from? The 24-hour day seems also to have originated in Egypt. From around 2400 BC the Egyptians began to tell the time at night in hours, according to the rising of the stars. Our evidence comes from an odd source: coffin lids. 'Star clocks' – diagrams showing sequences of stars through the night and the year – were sometimes painted on the inside of these lids. Their primary purpose was to help the dead in the afterlife: only they could 'see' these images once the coffins were closed. But they indicate that the living were also using the stars as a way of dividing up the time of darkness. The coffin lids show us that the Egyptians watched a sequence of 36 stars as they rose just before dawn, at ten-day intervals through the year. Sirius was one of these stars. The precise identification of the other 35 is still debated. Collectively they are known as the decans, after the Greek name for the ten-day interval between their risings (*deka* being Greek for ten). The interval between the rising of one star

Early Egyptian water clock, 1415–1380 BC. Scales on the walls marked the hours as the water drained away. SCIENCE & SOCIETY PICTURE LIBRARY

and the next was an Egyptian 'hour'. The night was broken up into 12 such hours, and these 12 night-time divisions were, it seems, transferred to the period of daylight.

Another means of telling the time at night, also developed by the Egyptians but perhaps borrowed by them from Babylonia, was the water clock. The earliest examples date from about 1500 BC. The water clock was simply a bucket with a hole near the base to act as the outflow. We call this device by its Greek name *klepsydra*, which means water-thief, because water in the bucket is 'stolen' through the night as it drains away. Inside the bucket the walls were marked with measuring scales, rather like our modern buckets, except these were not to indicate the quantity of water in the bucket but to mark the hours. There was one scale for each month of the year, and they were adjusted to allow for the longer and shorter hours of the seasons.

I say longer and shorter hours because, for most purposes, unequal hours were the norm. Each day or night would be divided into 12 from sunrise to sunset, and 12 from sunset to sunrise. Since the length of daytime and night-time changes with the seasons, the length of each hour changed also. Summer daytime hours were longer, and winter daytime hours shorter. Only at the spring and autumn equinoxes would the hours be equal through the whole 24-hour period. This gives us what is called the equinoctial hour. We are used to the equal, equinoctial hour because clocks have forced it on us since their invention in the Middle Ages. However, in antiquity only astronomers tended to use it. In Upper Egypt, the difference between a summer hour and a winter one would be small, a matter of a few minutes, but once the hours were picked up by people further north, the differences became more marked.

The sundial may have found its way into Greece just before 500 BC, but there is nothing to suggest it was used by ordinary people. We know that in Athens over the next two centuries the water clock remained the mechanism of choice for marking time limits of speeches in law courts. And well into the fourth century BC much simpler means of telling the time were used. Characters in Greek comedy talk of telling when it's dinner time from the length of a hungry person's shadow. The idea of using the human body as a sundial is also neatly captured in a quip by the Roman emperor Trajan:

If you put your nose facing the sun and open your mouth wide, you'll show all the passersby the time of day.

Here the image of a shadow cast by someone's nose across the gaping hole of their mouth reminds us of the hollowed-out, hemispherical stone sundials that have survived from the third century BC. The earliest surviving Greek sundial comes from the island of Delos, where 25 sundials have been excavated. Another 35 have been unearthed in Pompeii, which was buried by the eruption of Vesuvius in 79 AD. Together, these sites provide a good sample of the various kinds of sundial available to the Greeks and Romans – in particular, the hemispherical and the planar. The hemispherical sundial is the most labour-intensive to construct, but the simplest to mark out because it captures the celestial dome on a matching concave, hemispherical surface. Its pointer hangs out over the hollow hemisphere. The flat-plane type of sundial is the easiest to construct, but the most difficult to mark out because it involves transferring the hemispherical dome of the sky on to a flat surface. A shadow tracking the movement of the sun through the year is cast by a pointer, which usually sits upright from the sundial's surface.

Typically, these sundials have two lines which mark the limits of the shadow cast by the pointer. These are the shadows of the winter and summer solstices. In between there will often be a third line marking the two equinoxes. Other lines may mark notable days in the year such as festivals or even, on some Roman sundials, the birthday of the current emperor. Criss-crossing these lines will be another series of lines which mark out the 12 hours of the day, so the whole surface of the dial looks as if it is covered by a spider's web, and indeed 'spider's web' was the name given to one type of sundial by the Greeks and Romans.

It is possible to check the accuracy of a sundial for its particular locality, because the accurate placement of the equinox and solstice lines is a function of the specific latitude in which the dial is to be set. The lines on the sundials from Delos show that most were designed to be used right there. By contrast, the sundials from Pompeii show far less accuracy and don't particularly suit the latitude: the days of the solstices and equinoxes are wrongly marked. But the daily hour lines match reality reasonably well, so it may be that among the Pompeians there was more emphasis on the time of day and on business within each day than on the time of year.

The Romans were able to tell a good story against themselves when it came to living with inaccurate sundials. When they captured the city of Catania in Sicily in 263 BC, they took back to Rome as part of the booty a Greek sundial which had been made for the latitude of that city – 37 degrees north. Even though the hour lines on the dial were known to be inaccurate for Rome, the Romans followed the time of the dial for 99 years, before replacing it with a more accurately marked one in 164 BC. We don't know how the Romans knew the sundial was inaccurate. The first water clock wasn't set up there until 159 BC, so they don't seem to have relied on another type of instrument to tell them the correct time of day.

The hour lines on dials were sometimes numbered. If this were done in Greek, the letters of the alphabet were used for numbers: Alpha for 1, Beta for 2, Gamma for 3, and so on. A Greek epigram tells us, 'Six hours are sufficient for work. But the rest, when set out in letters, say "Live!" to mortals.' This is a simple pun on the Greek letters for the hours 7, 8, 9 and 10 – Z H Θ I – which together mean 'live'.

By the second century AD, Rome was reportedly stuffed with sundials. The captivating instruments were everywhere – in public squares, temples, town houses and country villas. They were also used the length and breadth of the Roman world, from Spain to Greece and from Africa to Germany. There were even miniaturised portable versions, the ancient equivalent of the modern watch. The earliest still extant hails from Herculaneum, a town near Pompeii which was also destroyed by the eruption of Vesuvius. The dial is known as the Ham Dial because its distorted bronze plate looks just like a small leg of ham. A spike on one side was intended to throw a shadow on to a series of criss-crossing lines on the plate, from which you could read the hour of the day. Other portable dials are regularly circular or cylindrical, and some come with extra plates to suit different latitudes. In this way they foreshadow the astrolabe, the portable timepiece *par excellence* of the Middle Ages.

Now, it's one thing to mark out the passage of time from hour to hour or day to day, but how did this fit into a vision of years? This was a very complicated issue, partly because most ancient peoples didn't measure the year by the sun, as we do, but by the moon. The sun gives us a year of 365.25 days on average, and we have used this kind of year ever since Julius Caesar imposed it on his fellow Romans in 46 BC. It's been tinkered with only twice since. The first tinkering happened just a few years after the solar year was introduced. The Romans had by mistake added the leap day every third year, instead of every fourth. The emperor Augustus fixed this problem in 9 BC by decreeing there would be no leap days for the next 12 years so as to allow the calendar and the sun to get back into sync with each other.

The second repair was made in 1582 by Pope Gregory XIII. The festival of Easter has always been partly tied to the sun, specifically to the spring equinox, and by the sixteenth century it had become increasingly obvious that Easter was being celebrated at the wrong time because the calendar was running ten days out of sync with the sun: while the calendar said the spring equinox took place on 21 March, in reality it was occurring ten days earlier, on 11 March. The fix this time was to chop the ten days out of the current year, and then avoid a leap day in those century years, such as 1700, 1800 and 1900, whose first two digits were not divisible by four. Because this was a change brought on by the leader of the Catholic church, it took a long time for Protestant Europe and America to accept it was necessary. Britain and her colonies didn't accept the revised calendar until 1752, by which time they had to chop off eleven days in that year. Other countries and cultures left change until the twentieth century, or avoided it entirely, which is why the Greek and Russian Orthodox churches usually celebrate Easter at a different time from the Western Christian churches: they are still running on Julius Caesar's calendar rather than Pope Gregory's. It's also why Russian athletes arrived 12 days late for the 1908 London Olympics: back then the Russian state was also still running on the Julian calendar.

Interestingly, the main calendar in ancient Egypt was very nearly a solar one, since it had exactly 365 days. It consisted of 12 months, each with 30 days. Five extra days were added at the end of each year to bring the total up to 365. We don't know for certain why the Egyptians took 365 days as the total for a year, but it may have had something to do with two natural events: the annual flood of the Nile, and the almost coincident dawn rising of the star Sirius. At some stage the Egyptians noticed that the first morning rising of Sirius more or less coincided with the start of the Nile flood. The earliest evidence for this association is on an ivory tablet from the ancient city of Abydos in the first dynasty (about 2925–2715 BC) in which Sirius is described as 'Bringer of the New Year and of the Inundation'. Over a long period of time the Egyptians could have seen that the average time between the dawn risings of Sirius, and between the floods of the Nile, was about 365 days.

A year of 365 days is close enough to the true solar year to make the Egyptian calendar an extraordinary achievement when compared with what the country's neighbours were doing in the Middle East, where the moon and not the sun ruled the calendar. But in the long run even 365 days is not long enough to avoid a displacement between this calendar and the seasonal year of 365.25 days, and so between religious and agricultural festivals and the relevant seasons. Every four years, the Egyptian calendar ran a day ahead of the sun; over a century it ran 25 days too fast. Only after 1461 years did it return to its original relationship with the sun. This long period was known as the Sothic year, after the Egyptian name – Sothis – for the star Sirius, whose dawn rising was meant to control the start of each year but which in reality failed to do so. The Egyptians themselves were aware of this drift but did nothing to stop it. The closest they came to a correction was in 238 BC, when King Ptolemy III decreed that an extra day should be added to the calendar every fourth year in order to slow it down. Despite the decree, the leap-year rule was never put into effect and the year continued to wander against the seasons. It seems that the fixed length of the Egyptian year had a strong symbolic significance which would brook no alteration.

IF THE STARS AND THE SUN can give us calendars, what about the moon? Well, the moon's regular phases led ancient peoples to use the month – the period of time between new moons – as a basic unit of time long before they thought of using the sun. A

The Unconquered Sun

month has advantages over a solar year, which is too long as a single unit of measurement for practical everyday use. Although one side-effect of the take-over of the Roman calendar was to put an end completely to any connection between the moon and the month, we still have moon calendars – at least in the sense that we are familiar with the idea of planting or fishing by the moon. In fact the Romans left us with an amazing amount of folklore about the moon's influences on farming. We are told, for instance, that crops should generally be planted just before the moon begins to wax, or during the waxing period. In other words, as the moon grows so too will the plants. On the same basis, harvesting should take place during the waning moon.

Variations are allowed, depending on the nature of the end-product. Grapes, for instance, may be picked under the waning moon if they are to be dried, but should be picked under the waxing moon if they are meant for making wine. But beware: the time for having your hair cut is also governed by the moon, and you should avoid having it cut at the time of the waning moon, for fear of going bald.

To get back to the lunar calendar, a lunar year usually consists of 12 months, or 354 days on average. The trouble with such a year is that it doesn't dovetail at all well with the seasonal year, which is ruled by the sun and consists of about 365.25 days. The effect of the difference is well illustrated nowadays by the wanderings of the Islamic religious calendar. This is a lunar calendar. A lunar month is not a whole number of days, but instead about 29.5 days on average. To allow for this, the months of the Islamic year are usually made up alternately of 29 and 30 days, for a total of 354 days. Hence, the lunar year falls short of a solar year by about eleven days, and a new lunar year starts before the old solar year is finished. The Islamic calendar doesn't try to realign the two types of year, so the Islamic religious year drifts through the seasons by about eleven days every solar year. We can see the effect of this in the movement of the holy month of Ramadan from year to year. Over a period of solar years, Ramadan will run through each of the seasons, at one time occurring in winter, then progressively in autumn, summer and spring. The whole cycle takes about 33 years to bring Ramadan back to the original point in the solar year.

The problem for societies which run both lunar and solar systems of reckoning time is how to make the two types of year equal. It's impossible to have a whole number of lunar months in a single solar year: a solar year consists of more than 12 but less than 13 lunar months. What people discovered early on, however, is that for practical purposes a certain number of solar years can contain a whole number of lunar months. We call these mixed systems 'lunisolar'. Most years will require only 12 lunar months, but an occasional one will need to have a 13th month added.

The earliest evidence of people realising they needed extra lunar months comes from Babylon. The Babylonian calendar stretches back to around 2000 BC. The names of the months had originally been associated with agricultural activities such as 'Ploughing' and 'Sheep', which were to be undertaken during them. So keeping the lunar calendar in sync with the sun was probably quite important, and there is very early evidence of such synchronisation. For instance, Hammurabi, king of Babylon from 1848 to 1806 BC, once decreed that a particular year was to be a leap year, in which the month Ululu would be repeated and called 'the second Ululu'. Tax that was meant to be paid in the following month, Tashritu, would now have to be paid in the second Ululu.

The Unconquered Sun

For a long time an extra lunar month was added at irregular intervals. It was not until the fifth or fourth century BC that a regular cycle was imposed on the lunar calendar. At that stage the Persians, who by then controlled Babylon and had adopted its calendar, introduced a cycle of 19 lunar years, with an extra month added in seven of the years – 3, 6, 8, 11, 14, 17 and 19. This works in the short term because 19 lunar years plus seven extra months equals 235 lunar months, which equal 6940 days, which also equal 19 solar years, more or less.

The Greeks also had systems for synchronising the lunar and solar years. They discovered the 19-year cycle in the fifth century BC, but they also used an eight-year cycle in which eight lunar years had three extra months added at various intervals. By the end of the cycle the sun and moon were pretty well in the same relationship as when the cycle started. There was, though, enough slippage in the cycle for the lunar calendar to run ahead of the sun by more than 1.5 days at the end of eight years. This may sound small, but after just nine eight-year cycles, or 72 years, the difference would amount to 14 days. So within a good lifetime, the calendar would have slipped the distance between a new moon and a full one, and obviously this could affect the celebration of a religious festival attached to, say, the full moon.

Even the 19-year cycle couldn't entirely remove the slippage, because over a period of four such cycles, or 76 years, the lunar calendar would still run ahead of the sun by a day. One solution was to remove that extra day after four cycles. This method was used in the Middle Ages and called 'the leap of the moon'. By then the 19-year cycle was being used to calculate the date of Easter.

Today Easter remains, for us in the Western tradition, a rare instance where we still have to struggle with the difference between the lunar and solar year. Easter Sunday is defined, generally, as the first Sunday following the fourteenth day after the first new moon (a lunar event) after the spring equinox (a solar event). This, in turn, is based on the definition of the Jewish festival of Passover, in the period of which the first Easter occurred. Easter contrasts with the other major Christian festival of Christmas, which is attached firmly to one day of the solar year, 25 December – the same date, incidentally, as the midwinter solstice in antiquity and the Romans' Birthday of the Unconquered Sun.

It is also not unusual now to find different New Year's days, such as the Western and Chinese, being celebrated in one country. This, too, is because the Chinese calendar is partly based on the moon and so its New Year's day is mobile. Despite the confusion, we continue to live with these two almost incompatible systems of measuring time, the solar and the lunar, for reasons drawn more from metaphysics than physics: lunisolar calendars, like the Christian, Jewish and Chinese, have strong theological or philosophical underpinnings which do not allow change.

Underneath the variety of means of keeping time, though, is a common principle: using the apparent paths of celestial bodies. The apparent movement of the sun around the Earth – or, as we now realise, the Earth's rotation about its own axis – has successfully defined time for millennia. The ordinary civil day comprises the interval between one noon and the next, between successive moments when the sun is at its highest in the sky. However, it has become apparent that the Earth does not spin uniformly, but is both slowing down and erratic. It is slowing down because of the frictional tidal effects of the moon on the Earth's oceans, and it is erratic because of the displacement of the North and South Poles by a few metres from one year to the next. In addition, seasonal fluctuations due to the varying distribution of air and water across the surface of the Earth over time cause the Earth to slow down in spring and speed up in winter.

Modern scientific theory and practice demand uniform time to ever-increasing levels of precision. Averaging out the days to produce a 'mean solar day' allows us to smooth out some of the wrinkles inherent in measuring time by the rotation of the Earth, but not to the degree of precision demanded: as we work at the subatomic level, smaller and smaller fractions of a second are needed. Until 1956 the 'second' was simply one 86,000th of a 'mean solar day' because there were 86,000 seconds in a day. From 1956, in an effort to provide greater standardisation, the 'mean solar second' was anchored artificially to the value it had in 1900. This continued to prove unsatisfactory, and so from 1967 it was agreed we should cut the conceptual umbilical cord to the rotation of the Earth, and measure time according to another system entirely – the natural vibrations of the caesium atom, which occur in the microwave part of the radio spectrum. The 'second' is now a measure of the frequency of the radiation emitted by a caesium atom, and is defined as the elapsed time of 9,192,631,770 oscillations of the 'undisturbed' caesium atom.

Atomic clocks (which have no face or arms) are based on these oscillations, much as older clocks were based on the oscillations, or swings, of a pendulum, but now there is very little to cause the effective 'pendulum' of the atom to wear down and lose time. Other atomic particles have been used or experimented with – including the ammonia molecule (a mixture of hydrogen and nitrogen, whose very combination caused problems of stability), and the rubidium atom (which is of lower quality than the caesium atom, but also relatively cheaper). But caesium atomic clocks provide an astonishing accuracy, losing only one second in three million years.

An even greater precision can be achieved with hydrogen maser clocks. 'Maser' is an acronym for 'Microwave Amplification by Stimulated Emission of Radiation', and these clocks are based on the radiation emitted by the hydrogen atom alone, which is still in the microwave region. Maser clocks can reach an accuracy 100 times better than that of caesium clocks, but only over short periods of about a week because of an instability caused by the collisions between the hydrogen atoms and the quartz bulb through which they pass.

Since 1972 the increased precision in measuring time has led to the addition of a 'leap second' every few years (most recently at the start of 2006) to bring our traditional method of timekeeping by the sun – which we retain in ordinary life – back into line with what our more refined clocks tell us. But while officially the basis for our timekeeping has shifted from the sun to the atom, and we understand the mechanics of the universe much better, in our daily lives our perspective hasn't changed. Not only is an atomic second still a second, but for us the sun continues to rise.

The Unconquered Sun

Robert Hannah *is Professor in Classics at the University of Otago. After receiving a Bachelor of Arts at Otago, he studied classical archaeology at the University of Oxford and in 1980 joined the staff of Otago's Classics Department. He has published widely on ancient Greek and Roman art and archaeology, and in recent years has focused on the cultural uses of astronomy in the classical world. He is the author of* Greek and Roman Calendars *(Duckworth, London, 2005) and is currently writing* Time in Antiquity, *to be published by Routledge.*

Galileo's Dilemma
—*John Stenhouse*

'SCIENCE WITHOUT RELIGION IS LAME, religion without science is blind,' Albert Einstein declared in 1940. The twentieth century's greatest physicist, writing at the beginning of World War II, believed it was imperative to maintain a civil and creative conversation between the world's scientific and religious traditions: respectful dialogue would, he believed, enhance both traditions – and the common good of humanity.

Today, Einstein's vision seems still pertinent. Modern science has spread around the globe. Hospitals from Turin to Teheran offer the latest scientific medicine. Physicists from Berlin to Beijing discuss the mysteries of the quantum world. But nations have not always used their scientific knowledge and power in unequivocally wise and humane ways. In 1945, for example, the United States, having enlisted expert physicists into its war effort, dropped atomic bombs on Hiroshima and Nagasaki. The bombs incinerated tens of thousands of Japanese people and ushered the world into the nuclear age. Today, many countries possess nuclear weapons, and some have sold and will sell the know-how, technology and even the uranium to others. Meanwhile, traditional religions that some scholars predicted in the 1970s would soon be extinct have roared back into public prominence. Christianity, while contracting in the West, has boomed in Africa, Latin America and parts of Asia, becoming once again a predominantly non-Western religion. Even in the relatively secular West, Christian traditions have by no means disappeared, as American political rhetoric illustrates. And no one can doubt the power of Islam.

If scientific and religious traditions are destined to remain permanent features of the twenty-first century world, the interesting question becomes what shape these traditions will take. Will they, as Einstein hoped, engage in civil, courteous and creative conversation? Or will cultural wars escalate? How do we build a world in which scientific traditions flourish while individuals and communities peacefully pursue religious and secular visions of the good life?

In 1981, Pope John Paul II – who, like Polish astronomer Nicholas Copernicus four centuries before, had studied at Crakow's Jagiellonian University – established a commission to review the Galileo case. In its report, issued in 1992, the commission acknowledged that the Roman Catholic church had made serious mistakes in condemning Galileo in 1633. The same thing could happen again, the Pope warned, unless church authorities treated scientific disciplines with the respect they deserved.

By revisiting the Galileo affair, the Pope also hoped to criticise what he called a 'sort of myth' among many modern thinkers that science and faith were incompatible. I want to reflect on this claim in light of recent historical scholarship. Have myths and half-truths about the past really circulated in our allegedly sceptical, rational, no-nonsense modern world?

First, though, some background. When we think about science and religion we tend to assume there were two distinct entities in the past to which these labels applied. This is not strictly true. Until the beginning of the nineteenth century, science and religion often overlapped and interpenetrated. Many

Galileo's Dilemma

of the greatest scientific minds of sixteenth and seventeenth century Europe – Copernicus, Johannes Kepler, Galileo, Francis Bacon, Isaac Newton and Robert Boyle – often did not sharply distinguish their science from their religion. Newton, the great English natural philosopher who revolutionised physics and cosmology during the late seventeenth century, also embraced an idiosyncratic form of Christianity, which profoundly affected the way he thought about nature. In his most famous book, *Principia Mathematica*, published in 1687, he described the workings of the solar system, with planets and satellites all moving in the same direction and in the same plane while comets coursed among them. 'This most beautiful system of the sun, planets and comets,' he wrote, 'could only proceed from the counsel and dominion of an intelligent and powerful Being.'

Newton's theistic science leads to my second point, that historians of science try to understand historical figures in terms of the scientific and religious ideas and values prevailing in *their* times, not ours. Take James Ussher, a seventeenth century Anglican archbishop of Armagh. Working back through the genealogies in the Bible, Ussher calculated that God had created the world in 4004 BC. It's hard not to smile, but given the science and biblical scholarship of his time, Ussher's dating made reasonable sense. He was an able and intelligent scholar.

Thirdly, the terms science and religion have come under scrutiny in recent years. We often speak of what 'science' states, or 'religion' claims, as though science and religion were existing things, capable of speaking or claiming. In fact, it is human beings who do the stating and the claiming, often in the name of 'science', 'nature', 'God' or 'history', and with them come mundane human interests and agendas. Past science–religion

Italian astronomer Galileo Galilei supported Copernicus's theory that the sun, not the Earth, was the centre of the solar system. Under pressure from the Roman Catholic Inquisition, he recanted in 1633 but was held under house arrest for the rest of his life. SCIENCE, INDUSTRY & BUSINESS LIBRARY/NEW YORK PUBLIC LIBRARY/SPL

encounters occurred not just between abstract systems of ideas, but between flesh-and-blood human beings acting in concrete social and cultural contexts. Issues of personality, authority and power often profoundly shaped ostensibly high-minded debates about science and religion.

With these points in mind, let's turn now to seventeenth century Italy. In 1633 the Roman Catholic Inquisition, with the backing of Pope Urban VIII, tried Galileo Galilei, the most famous Italian scientist of the day, on the charge of publicly defending Nicholas Copernicus's theory that the sun, not the Earth, lay at the centre of the solar system. Galileo argued that the Earth, a planet like Mars or Mercury, spun on its axis once a day and revolved annually around the sun. His opponents, following the older physics and astronomy of Aristotle and Ptolemy, believed that the heavenly bodies – the planets, sun

Galileo's Dilemma

Galileo was the first to use a telescope for astronomical observations, after the device was invented in Holland. Here he is depicted in St Mark's Square, Venice in 1609, standing to the right of the telescope, which is being examined by the Doge of Venice. SPL

and stars – orbited a motionless Earth. Certain passages in the Psalms and the book of Joshua, read literally, appeared to confirm Earth-centred cosmology.

Galileo's inquisitors showed that he had publicly defended the Copernican system not just as a useful way of calculating the movements of the planets, but as physically true. This, he said, was how the solar system actually worked. Some years earlier Galileo had promised leading Catholic theologian Cardinal Roberto Bellarmine he would not do this. Threatened with torture if he refused to recant, 69-year-old Galileo faced a dilemma. Should he renounce his scientific convictions? Or should he stick to his guns, defy the Inquisition, and take the consequences? Kneeling in a white shirt, Galileo renounced his support for Copernicus. According to one story, probably apocryphal, he muttered, under his breath, 'Eppur si muove'

Charles Darwin, the British naturalist whose publication in 1859 of *The Origin of Species* saw the beginning of a fierce debate over evolution which still surfaces two centuries later.

('Yet it does move'), recanting his recantation. The church authorities sentenced him to house arrest and required him to repeat penitential psalms for three years. His elder daughter Suor Marie Celeste, a nun in the cloistered order of the Poor Clares, recited the penitential psalms on his behalf.

Galileo's trial has echoed down the centuries. Ask people about science and religion and this is still the incident most likely to spring to mind. But it was far from the first. In 1492, for example, Christopher Columbus, setting sail from Spain for the Indies, allegedly had to defy a chorus of Catholic theologians who, convinced the Earth was flat, warned him he would sail off the edge, probably into hell.

One of the famous incidents of the nineteenth century was the so-called battle over evolution which climaxed at an Oxford meeting of the British Association for the Advancement of

Galileo's Dilemma

LEFT: Samuel Wilberforce, Bishop of Oxford, an early critic of Charles Darwin's theory of evolution, met his match at a famous 1860 meeting of the British Association for the Advancement of Science. GEORGE BERNARD/SPL

RIGHT: Darwin supporter T. H. Huxley, as portrayed in an 1881 edition of *Punch*. Huxley was so strident in public meetings he became known as 'Darwin's bulldog'. GEORGE BERNARD/SPL

Science. Anglican Bishop 'Soapy Sam' Wilberforce concluded a critique of Darwin's *The Origin of Species*, published the previous year, by asking biologist T. H. Huxley, an ardent defender of Darwin's ideas, whether it was through his grandfather or grandmother that he claimed to be descended from a monkey. Huxley replied that he would rather be descended from an honest ape than from a bishop who abused his talents and high position in the service of falsehood and religious prejudice – thus launching into a long and spectacular career as the world's first self-styled episcopophage, or bishop-eater.

Across the Atlantic, fundamentalist Protestants battled evolution and modernism. At Dayton, Tennessee in 1925, Presbyterian layman William Jennings Bryan, a three-time Democratic nominee for president, led the team that prosecuted high-school teacher John T. Scopes for defying a state law that banned the teaching of evolution in schools. Bryan and the anti-evolutionists won the court battle, but in the eyes of their critics it was a pyrrhic victory. The journalist H. L. Mencken, who covered the trial for *The Baltimore Sun*, gave thanks that Calvin Coolidge and not William Bryan governed America. 'The president of the United States may be an ass,' he wrote, 'but at least he doesn't believe that the Earth is square, that witches should be put to death, and that Jonah swallowed the whale.'

Historians, philosophers and popular writers have regularly cited incidents such as these to argue that organised religion has repeatedly tried to suppress or discredit new scientific theories and their proponents. Let's call this view of history the warfare thesis. Its proponents typically depict science and religion as locked in perpetual conflict, with church authorities and theologians generally suspicious of, or hostile towards, science. Churchmen such as Pope Urban VIII, Bishop Wilberforce and William Bryan appear as the villains, harassing or persecuting the good guys such as Copernicus, Galileo and Darwin. Despite trials and tribulations, science triumphs in the end.

The warfare thesis first appeared on the intellectual map of Europe during the eighteenth century, when Enlightenment writers such as Voltaire and Condorcet, who hated the powerful European state churches primarily for moral and political reasons, set out to smash them with science, history, philosophy and wit. In the nineteenth century, two American historians gave the warfare thesis the appearance of sound scholarship. In 1874, New York chemist-turned-historian John W. Draper, a lapsed Methodist, attacked the Roman Catholic church for what he claimed was its perennial hostility toward science.

His *History of the Conflict between Religion and Science* was a bestseller, running through 50 printings in Britain and 21 in the United States, and being translated into ten languages. Draper argued that the Vatican's relentless persecution of scientists such as Galileo had left its hands 'steeped in blood'. His anti-Catholicism pleased many Protestant and secular readers angry at the Vatican's refusal to embrace modern liberal and progressive thinking.

Two decades later, Episcopalian historian Andrew Dickson White, president of Cornell University, extended Draper's thesis to the Protestant churches. In his 1896 book *A History of the Warfare of Science with Theology in Christendom*, he depicted history as a series of battles between narrow-minded, dogmatic Christian theologians on one side, and truth-seeking, open-minded men of science on the other. White had a reason to feel sour toward Protestant clergy. During the 1870s he had led a campaign to win a land grant from the New York State Senate to establish Cornell as the nation's first openly secular research university. After he succeeded, his Protestant rivals, bothered by the precedent, had set about publicly criticising him and his 'godless' university.

White's two-volume history, which included massive footnotes, had a wider and more lasting impact than Draper's book. The rationalist philosopher Bertrand Russell acknowledged relying on it in writing his own book, *Religion and Science*, in 1935. As late as the 1960s the work was still being reprinted and praised by historians. In New Zealand, Lloyd Geering, the country's most prominent religious thinker, enthusiastically embraced the Draper–White view.

However, despite its long popularity, in recent years the

153

warfare thesis has taken such a pounding that professional historians of science have largely abandoned it. This is not to defend the opposite view, that Christianity and science have had an invariably peaceful, harmonious and productive relationship. Let's call this the harmony thesis. Some of its proponents, understandably annoyed by the excesses of the warfare thesis, have argued that, far from stifling science, Christianity provided the vital ingredients for its birth. How, they ask, could the scientific revolution in Europe have occurred during the period of heightened religiosity that followed the Reformation if Protestants and Catholics alike had tried to suppress scientific thinking and practice? It's a fair question, and illustrates the problems facing the warfare thesis.

Yet the harmony thesis, too, can be overstated. Some enthusiastic harmonists have given Christianity virtually all the credit for the rapid development of the sciences in early modern Europe, and failed to adequately acknowledge the scientific achievements of the Greek, Islamic, Chinese and Indian civilisations, from all of which European Christians borrowed extensively. Moreover, the harmony thesis ignores the fact that Christians have sometimes obstructed free inquiry, harassed scientists and said silly things about science. In his 1992 review of the Galileo affair, Pope John Paul II acknowledged as much.

Let's return now to 1492, with Columbus about to set sail for the Indies while credulous medieval Catholics try to terrify him out of embarking. Versions of this flat-Earth story were being peddled in American university and high-school history textbooks as recently as the 1980s. It is exciting, dramatic, widely believed – and almost entirely false. As historian Jeffrey Burton Russell showed in his book *Inventing the Flat Earth*, the real error 'is not the alleged medieval belief that the Earth was flat, but rather the modern error that such a belief ever prevailed'. Virtually all educated Christians during the high Middle Ages knew the Earth was round; the ignorant medieval Catholic is largely a product of Protestant and Enlightenment prejudice.

What about Galileo, often presented as a classic case of a brilliant, heroic scientist crushed by a reactionary, authoritarian church? Galileo was certainly a brilliant and original scientific thinker, many of whose ideas turned out to be correct. But scientific evidence in favour of a sun-centred solar system was not clearcut in 1633 and Galileo's crowning proof – that the rise and fall of the tides proved the Earth was moving – not only convinced few at the time but turned out to be mistaken. Moreover, although Galileo had early supporters among the clergy, particularly Jesuit astronomers, he managed to alienate most of them, even managing to insult Pope Urban VIII, an urbane and cultured man with whom he had been friendly when the priest was Cardinal Maffeo Barberini. By assailing, sometimes savagely, those with whom he disagreed, Galileo was behaving very much as Italian Renaissance court culture expected: a fiery intellectual favourite who gave no quarter to his opponents reflected honour on his aristocratic patron. Unfortunately, he was to learn that a favourite could fall as swiftly as he had risen.

The wider cultural context must also be noted. After Copernicus published *On the Revolutions of the Heavenly Spheres* in 1543, heliocentrism had caused no stir for half a century. But after the Reformation religious and political divisions intensified, plunging Europe into the Thirty Years War of 1618–48. By 1633 the Roman Catholic hierarchy, having lost about half of western Europe to Protestantism, was taking a

JOHN STENHOUSE

much harder and more literal line on biblical interpretation than it had previously. As Galileo discovered, it was a particularly bad time to be debating science and scripture.

The Galileo affair cannot, then, be interpreted as a simple battle between science and religion. Everyone involved, whether a church official or a disciple of Galileo, called himself a Christian. Every significant participant, without exception, acknowledged the authority of the Bible. Virtually all, including Galileo, were theologically well-informed and articulated carefully reasoned theological positions. Virtually all held informed, rationally defensible, cosmological beliefs. Conflict erupted as much within the church and within science as between science and the church. The title of a recent book by Italian philosopher and astronomer Annibale Fantoli sums up Galileo's position: *For Copernicanism and For the Church*.

A leading science historian, Professor David Lindberg, has concluded that, although what happened was shocking by our standards, by the standards prevailing in seventeenth century Europe the central bureaucracy of the church and the people who staffed it lived up to widely held norms, followed accepted procedure, and even on a number of occasions treated Galileo with generosity. The Galileo affair, Lindberg argues in *When Science and Christianity Meet*, co-written with Ronald L. Numbers and published in 2003, was 'a product not of dogmatism or intolerance beyond the norm, but of a combination of more or less standard (for the seventeenth century) bureaucratic procedure, plausible (if ultimately flawed) political judgement, and a familiar array of human foibles and failings'.

Did the Galileo affair strangle science in Catholic Europe, and turn scientific development over the following centuries into a predominantly Protestant affair, as some historians have claimed? Undoubtedly church authorities did constrain what Catholic scientists could publish, but that did not stop talented scientists such as René Descartes, Marin Mersenne, Pierre Gassendi and Blaise Pascal making important contributions to a range of disciplines. In a remarkable book entitled *The Sun in the Church*, published in 1995, science historian John Heilbron described how the Catholic church, cultivating astronomy in order to refine the church calendar, turned European cathedrals into gigantic solar observatories. According to Heilbron, for over six centuries the church gave more financial and social support to the study of astronomy – from the recovery of ancient learning during the late Middle Ages to the Enlightenment – than any other institution. This conclusion, though counter-intuitive to those raised on the warfare thesis, has won wide assent among historians of science.

During the seventeenth and eighteenth centuries Catholic missionary orders, with Jesuits leading the way, exported the latest European science and technology – minus Copernicanism – to the wider world, and in turn brought back to Europe new knowledge of places, plants and peoples of Asia and the Americas. Global Catholic missionary networks constituted a kind of early worldwide web of science. Indeed, during the nineteenth century Protestant missionaries took a lead from their Catholic counterparts, sending back to Berlin, Paris, London, Oxford and Edinburgh information, maps and specimens of the plants, animals, languages, places and peoples they encountered. Many missionaries served as valued collectors and fieldworkers for metropolitan experts in Europe. The best won recognition as outstanding scientists and scholars.

JOHN STENHOUSE

What about Darwin and evolution during the nineteenth century? Here, surely, the warfare thesis holds? T. H. Huxley, Darwin's fiercest supporter, certainly thought so. 'Extinguished theologians,' he wrote in 1860, 'lie about the cradle of every science like the strangled snakes beside that of Hercules; and history records that, whenever science and [Christianity] have been fairly opposed, the latter has been forced to retire from the lists, bleeding and crushed if not annihilated.'

Yet Huxley's high-flown rhetoric, aimed at drumming amateurs and parsons out of science, cannot be regarded as a reliable guide to history. Huxley's biographer Adrian Desmond has pointed out that the biologist and educator, born above a butcher's shop, hated the Anglican establishment that dominated English science, education and politics of the time. A brilliant polemicist, he savaged the church in a largely successful campaign to take control, capitalising on powerful religio-political forces that were already transforming English culture. He succeeded in large part because he attracted the support of a large body of Nonconformist Protestants, whose campaign against Anglican domination predated his.

Huxley did not merely aim to drive Anglicans and Catholics out of professional scientific societies and institutions. He also campaigned, often behind the scenes, to exclude women, who in his view lacked the intelligence and rationality to engage in genuine scientific research and were too fond of churchgoing. Writing to geologist Charles Lyell, he declared that 'five-sixths of women will stop in the doll stage of evolution, to be the stronghold of parsondom'.

What about Bishop Wilberforce, depicted by supporters of Huxley as a sneering ignoramus? The truth here was also

Galileo's Dilemma

somewhat different. Richard Owen, Britain's leading comparative anatomist, had in fact primed the bishop with up-to-date knowledge of natural history. Wilberforce's scientific criticisms of *The Origin of Species* impressed Darwin himself. Although Huxley and his allies claimed victory at the British Association debate in 1860, contemporary opinion was divided and many considered the Huxley–Wilberforce exchange to have been a draw.

IN NEW ZEALAND, unlike England, there was no established church. Hence, local naturalists had little need to use evolution as a weapon with which to chase a domineering religious establishment out of science. F. W. Hutton, one of New Zealand's leading nineteenth century zoologists, illustrates the difference. While living in England as a 23-year-old geologist, Hutton had written one of the first-ever reviews of *The Origin of Species*, praising Darwin's theory as the best available to explain the origin of species, and attacking 'direct creation', in language reminiscent of Huxley, as 'verbal hocus-pocus' and 'a specious mask for our ignorance'. The review delighted Darwin.

Once he arrived in New Zealand, however, Hutton largely left such polemic behind. In the more open and fluid environment, he and leading naturalists such as James Hector, Julius von Haast, Walter Buller, G.M. Thompson and A.P.W. Thomas found men of science and church people more willing to listen. Indeed, in a series of lectures on evolution held at Otago University in 1876, Hutton managed to convert two of the city's leading clergy – Anglican Bishop Samuel Nevill and Presbyterian Professor of Divinity William Salmond – to evolutionary views. The fact that Hutton, a liberal Anglican, argued that God lay

JOHN STENHOUSE

Methodist minister Alfred Fitchett, whose exclusion from membership of the Dunedin YMCA because of his belief in evolution led to a fiery public meeting. THE HOCKEN LIBRARY

behind the natural laws and processes Darwin had identified almost certainly helped wavering Christians embrace evolutionary theory.

Similarly, in nineteenth century America, which also lacked a church establishment, most scientists embraced evolution without abandoning their religious beliefs. In his recent book *Darwinism Comes to America*, the science historian Ronald L. Numbers examined the religious views of 80 prominent American scientists – mostly geologists, biologists and anthropologists – elected to the National Academy of Sciences during the second half of the century. As far as he could discover, not a single one had substantially altered their world-view as a result of embracing evolutionary theory. The Christians had remained Christian, the agnostics agnostic.

New Zealand did not, however, completely escape conflict over evolution. As with the Galileo affair, the most explosive

Galileo's Dilemma

debate took place *within* the churches. Early in 1876 Alfred Robertson Fitchett, Dunedin's leading Methodist minister, accepted evolution in sermons and in a published pamphlet. Evidently familiar with Darwin's arguments in *The Origin of Species* and *The Descent of Man*, he dismissed the idea that God had miraculously created Adam and Eve as 'popular craving for creation by fiat' and proposed that thinking Christians come to terms with evolution by interpreting the Genesis creation accounts metaphorically, not literally.

In September that year, Fitchett applied to join the Dunedin branch of the Young Men's Christian Association, an interdenominational Protestant body, but the board of management rejected his application. The *Otago Daily Times* was outraged. If Fitchett went to hell for being an evolutionist, the newspaper declared, he would roast alongside several local clergy. When the YMCA responded by calling a special general meeting, 120 of Dunedin's leading citizens packed the hall. The first to speak, Robert Borrows, a physician and member of Fitchett's Trinity Wesleyan congregation, criticised the YMCA for rejecting a Methodist minister in good standing who had the respect of a large Dunedin congregation. How, he declared, could the board impose a particular interpretation of Genesis on members of the YMCA when Protestants had the right to interpret the Bible for themselves?

Up to this point Borrows stood on firm ground and had the meeting with him. His next suggestion, however, sealed Fitchett's fate. If the YMCA considered itself truly interdenominational, Borrows suggested, it should be prepared to enrol 'enlightened Roman Catholics...even if they call the Virgin Mary Queen of Heaven and say their prayers before an image, so long as they acknowledge her Son as King of Glory'. This

161

was too much for some Protestants at the meeting, who greeted Borrows's suggestion with shouts of 'No!' and prolonged hissing. Borrows almost incited a riot by retorting, 'I do think some of you require a little more evolution. My ancestors gave up that language ages ago.' Eventually, the chairman restored order.

With friends like Borrows championing his cause, Fitchett scarcely needed enemies. The YMCA debate went on for two more hours, with conservative Calvinist minister James Copland offering a reasoned critique of Fitchett's liberal theology. But the damage had been done. The meeting voted 50 to 31 to reject Fitchett's application.

The *Otago Daily Times*, a supreme exemplar of non-partisan journalism, condemned Fitchett's critics as a 'handful of bigots who have prostituted the sacred name' and regretted the injury done to religion by 'this miserable outrage'. A week later 43 members withdrew from the YMCA in protest at Fitchett's treatment. Methodists, Baptists, Congregationalists and Anglicans led the exodus. Not all agreed with Fitchett's views, but all refused to kowtow to the authority of Scottish Presbyterians who, according to one critic, were behaving like self-made popes. These English Nonconformists were prepared to pull down the YMCA about their ears rather than allow conservative Scots to dominate Dunedin's intellectual and religious life.

The Fitchett affair demonstrated, yet again, how significantly personalities and power struggles shape debates on science and religion. The personal and political dimensions in all these encounters reveal much the same complexity and ambiguity that we find in other areas of human experience. As an alternative to the tidy simplicities of the warfare and harmony theses, this complexity thesis may seem at first glance a disappointingly tame note on which to conclude. But in a twenty-first century world increasingly polarised between secular and religious forces, highlighting complexity and taking an ironic perspective may serve us well. The new history of science-and-religion extends sympathetic historical understanding as widely as possible – to religious conservatives as well as to secular progressives, to Pope Urban, Bishop Wilberforce and William Jennings Bryan as well as to Galileo, T. H. Huxley and Clarence Darrow. Today, when science-and-culture wars simmer in the West and tensions with the Islamic world burgeon, we may all profit from the kind of history that sets out to understand and humanise, rather than to caricature or demonise those with whom we disagree.

John Stenhouse *is Associate Professor in the History Department, University of Otago, where he teaches the history of science. Recent publications include* Disseminating Darwinism: The Role of Place, Race, Religion and Gender, *co-authored with Ronald L. Numbers (Cambridge University Press, 1999) and* Building God's Own Country: Historical Essays on Religions in New Zealand *(Otago University Press, 2004). His next book will be on the role of Christian missionaries in the making and spreading of modern Western science.*

Further Reading

INTRODUCTION

Annus Mirabilis: 1905, Albert Einstein and the Theory of Relativity,
   John Gribbin and Mary Gribbin: Chamberlain Bros, 2005
Autobiographical Notes, Albert Einstein: Open Court Publishing
   Company, 1979
E=mc²: A Biography of the World's Most Famous Equation,
   David Bodanis: Pan Books, 2001
Subtle is the Lord: The Science and the Life of Albert Einstein,
   Abraham Pais: Oxford University Press, 1982

A SHORT HISTORY OF THE UNIVERSE

The Alchemy of the Heavens: Searching for Meaning in the Milky Way,
   Ken Croswell: Anchor, 1996
Black Holes and Time Warps: Einstein's Outrageous Legacy,
   Kip S. Thorne: W. W. Norton & Company, 1995
The Black Hole at the Center of Our Galaxy, Fulvio Melia:
   Princeton University Press, 2003
*The Elegant Universe: Superstrings, Hidden Dimensions, and the Quest
   for the Ultimate Theory*, Brian Greene: Vintage, 2000
*The Strange Case of Mrs. Hudson's Cat: And Other Science Mysteries
   Solved by Sherlock Holmes*, Colin Bruce: Perseus, 1997
The Whole Shebang: A State-of-the-Universe(s) Report,
   Timothy Ferris: Phoenix, 1998

DISCOVERING THE AGE OF THE EARTH

Awesome Forces: The Natural Hazards that Threaten New Zealand,
   Geoff Hicks and Hamish Campbell, editors: Te Papa Press, 1998

FURTHER READING

The Dinosaur Hunters: A True Story of Scientific Rivalry and the Discovery of the Prehistoric World, Deborah Cadbury: Fourth Estate, 2001

The Earth: An Intimate History, Richard Fortey: Harper Collins, 2005

EINSTEIN AND THE ETERNAL RAILWAY CARRIAGE

Einstein: A Life, Denis Brian: John Wiley & Sons, 1996

Einstein's Heroes: Imagining the World through the Language of Mathematics, Robyn Arianhrod: University of Queensland Press, 2004

Einstein's Wife: Work and Marriage in the Lives of Five Great Twentieth-Century Women, Andrea Gabor: Viking, 1995

SCHRÖDINGER'S CAT

A Life of Erwin Schrödinger, Walter Moore: Cambridge University Press, 1994

Quantum Generations: A History of Physics in the Twentieth Century, Helge Kragh: Princeton University Press, 2002

Schrödinger: Life and Thought, Walter Moore: Cambridge University Press, 1992

The Conceptual Development of Quantum Mechanics, Max Jammer: Tomash Publishers, 1989

JOURNEY TO THE HEART OF MATTER

Boltzmann's Atom: The Great Debate that Launched a Revolution in Physics, David Lindley: Free Press, 2001

Einstein's Miraculous Year: Five Papers that Changed the Face of Physics, John Stachel, editor: Princeton University Press, 2005

Rutherford: Recollections of the Cambridge Days, Mark Oliphant: Elsevier, 1972

FURTHER READING

Rutherford: Scientist Supreme, John Campbell:
   AAS Publications, 1999
Rutherford: Simple Genius, David Wilson:
   MIT Press, 1983
The Heart of the Antarctic, Ernest Shackleton:
   Penguin, 2000

THE UNCONQUERED SUN

From Sundials to Atomic Clocks: Understanding Time and Frequency,
   James Jespersen and Jane Fitz-Randolph: Dover Publications, 2000
*Greek and Roman Calendars: Constructions of Time in the Classical
   World*, Robert Hannah: Duckworth Publishing, 2005
The Calendar: David Ewing Duncan: Fourth Estate, 1998
The Calendar: Measuring Time, Jacqueline de Bourgoing:
   Thames & Hudson, 2001
The Discovery of Time, Stuart McCready, editor: MQ Publications, 2001
The Story of Time, Kristen Lippincott, editor, with Umberto Eco,
   E.H. Gombrich and others: Merrell, 1999

GALILEO'S DILEMMA

Darwinism Comes to America, Ronald L. Numbers: Harvard University
   Press, 1998
Galileo: For Copernicanism and For the Church: Annibale Fantoli; George
   V. Coyne, translator: Vatican Observatory Publications, 1996
Reconstructing Nature: The Engagement of Science and Religion,
   John Brooke and Geoffrey Cantor: Oxford University Press, 1998
Science and Religion: Some Historical Interpretations,
   John Brooke: Cambridge University Press, 1991
When Science and Christianity Meet, David C. Lindberg,
   Ronald L. Numbers, editors: University of Chicago Press, 2003

Index

Adams, Chris, 41, 43
Aspect, Alain, 96
asteroid impact, 26–28
atomic clocks, 141–42
Avogadro, Amadeo, 104
Avogadro's number, 104, 107
Becquerel, Henri, 5
Bell, John, 95–96
Bernoulli, Daniel, 102, 104
Beyrich, Heinrich Ernst, 33
big bang, 13–14, 16, 20
black holes, 13
Bohr, Niels, 82–83, 90–93, 111–12
Boltzmann, Ludwig, 105–107
Born, Max, 86, 91–92
Boyle, Robert, 103
Brongniart, Alexandre, 31
Brownian motion, 6–7, 13, 106–107
calendars, 122, 126–27, 134–40
Cambrian period, 32, 43, 45
Carboniferous period, 30
Columbus, Christopher, 150, 154
comets
    impact, 26–28, 45
    Shoemaker-Levy, 26
Compton, Arthur, 79–80
Copernicus, Nicholas, 148, 155
cosmic background radiation
    (CBR), 18–22

cosmic microwave background
    (CMB), *see* cosmic background
    radiation (CBR)
Cretaceous period, 31, 45
Curie, Marie, 5, 37–38, 40, 67
d'Halloy, Jean-Baptiste-Julien
    d'Omalius, 31
Dalton, John, 104
Darwin, Charles, 31, 37, 150–51,
    158–61
dating, *see* geochronology and
    radiometric dating
de Broglie, Louis, 83–85
Democritus, 102
Devonian period, 33
dinosaurs, 27, 34–37, 45
    extinction of, 27, 35, 37, 45
    Iguanodon (tooth), 34–36, 45
Draper, John W., 152–53
$E=mc^2$, 7–9, 39–40, 66
earthquakes, 32, 41
Einstein, Albert
    in America, 9, 55
    annus mirabilis (1905), 1, 6–10, 51
    in Bern, 4–9, 51, 54–55
    birth, 49, 51
    childhood, 1–2, 51–53
    children, 4, 50, 54
    death, 1, 49

167

INDEX

education, 2–3, 51–54
marriage, 4, 50, 55–56
Nobel Prize, 6, 49
papers, 6–8, 51
PhD, 6–7
on quantum mechanics, 91–95
entropy death, 25
entropy gradient, 26
Eocene epoch, 31
Euclidean geometry, 15–16
evolution, theory of, 37, 44, 150–51, 158–61
extinction, 27, 34–37, 45
Fitchett, Alfred Robertson, 161–62
fossils, 30–37
galaxy formation, 21–25
Galileo, 58, 146–50, 155–56
geochronology, 38–45
geological epochs
    Eocene, 31
    Miocene, 31
    Oligocene, 33
    Palaeocene, 33
    Pleistocene, 33
    Pliocene, 31
geological periods
    Cambrian, 32, 43, 45
    Carboniferous, 30
    Cretaceous, 31, 45
    Devonian, 33
    Jurassic, 31
    Ordovician, 33, 45
    Permian, 33, 45
    Silurian, 32
    Triassic, 32, 45
Heisenberg, Werner, 82–83, 88–90
Hubble flow, 16
Hutton, F. W., 159
Huxley, T. H., 151, 158–59
Iguanodon (tooth), 34–36, 45
Jurassic period, 31
Kirchoff, Gustav, 76
Lapworth, Charles, 33
Lavoisier, Antoine-Laurent, 103
Lenard, Philippe, 75, 77
Lightfoot, John, 29
light-year, 63–64
Lindberg, David, 156
Löwenthal, Elsa, 55–56
Lyell, Sir Charles, 31–33
Mach, Ernst, 106–107
Mantell, Gideon and Mary Ann, 34–36, 45
Marić, Mileva, 3–4, 50, 54–55
Marsden, Ernest, 100–101, 114–15
maser clocks, 142
mass spectrometer, 38, 40
meteorites, 45
Millikan, Robert, 78
Miocene epoch, 31

Moseley, Harry, 112, 114
Murchison, Sir Roderick Impey, 32–33
Nobel Prize
   Bohr, Niels, 82
   Einstein, Albert, 6, 49
   Heisenberg, Werner, 82
   Perrin, Jean, 107
   Rutherford, Ernest, 116
Oligocene epoch, 33
Ordovician period, 33, 45
Owen, Richard, 35
Palaeocene epoch, 33
parapegma, 127
Penzias, Arno, 20
Permian period, 33, 45
Perrin, Jean, 107
photoelectric effect, 13, 78
Planck, Max, 74–77, 112
Planck spectrum, 19
planetary formation, 21–25
plate tectonic theory, 45–46
Pleistocene epoch, 33
Pliocene epoch, 31
Podolsky, Boris, 94–95
Priestley, Joseph, 103
quantum mechanical tunnelling, 84, 87
quantum mechanics, 73–74, 82–95
quantum theory, 6, 73, 76–82

quark-gluon plasma, 16–17, 24
radioactive decay, 39–40
radiometric dating, 30, 38–39, 45
red giant, 25
relativity, 7-8, 13–16, 58–71
   general theory of, 13–16, 68–71
   special theory of, 7–8, 13, 62–67
Röntgen, Wilhelm Conrad, 5
Rosen, Nathan, 94–95
Rutherford, Ernest, 5, 38, 40, 67, 81, 99–101, 103, 108–16
Schimper, Wilhelm Philipp, 33
Schrödinger, Erwin, 83–84, 93
Schrödinger's cat, 93–94
Sedgwick, Adam, 32–33
sedimentary rock, 41–42
Silurian period, 32
Smith, William, 30
speed of light, 7–8, 59–66
star death, 25
star formation, 21–25
Stonehenge, 123–24
Stonehenge Aotearoa, 120
stratigraphy, 31–34, 37, 45
subduction, 46
sundials, 128–29, 131–34
supernova explosions, 22–25, 116
Thomson, George, 86
Thomson, J. J., 5, 75, 77, 109
timekeeping, 119–22, 126–42

169

INDEX

Triassic period, 32, 45
uncertainty principle, 82–83, 88–90, 95
Unconquered Sun, Birthday of the, 122
universe, formation, 13–25
    big bang, 13–14, 16, 20
    decoupling, 18–22
    recombination, 18
Ussher, James, 29–30, 147
volcanoes, 41, 45
von Alberti, Friedrich August, 32
water clock, 130–31
Wein, Wilhelm, 76
Wheeler, John, 15
White, Andrew Dickson, 153
Wilberforce, Samuel, 151, 158–59
Wilson, Robert, 20